Samsung Galaxy A71 User Guide

A Newbie Guide to Master the New Samsung Galaxy A71 like a Pro

Douglas A. Harting

Copyright © 2020 Sam J. Wealth

Introduction

The Galaxy A71 is an amazing devices. It comes with a big screen, incredible pictures, fresh functionality, and excellent specs to give you the most experience.

Device Layout

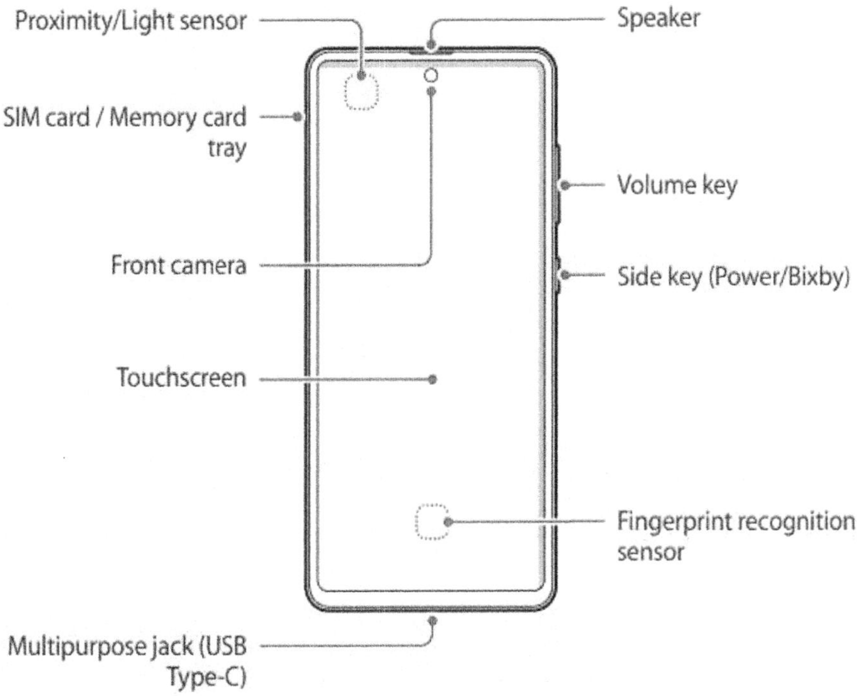

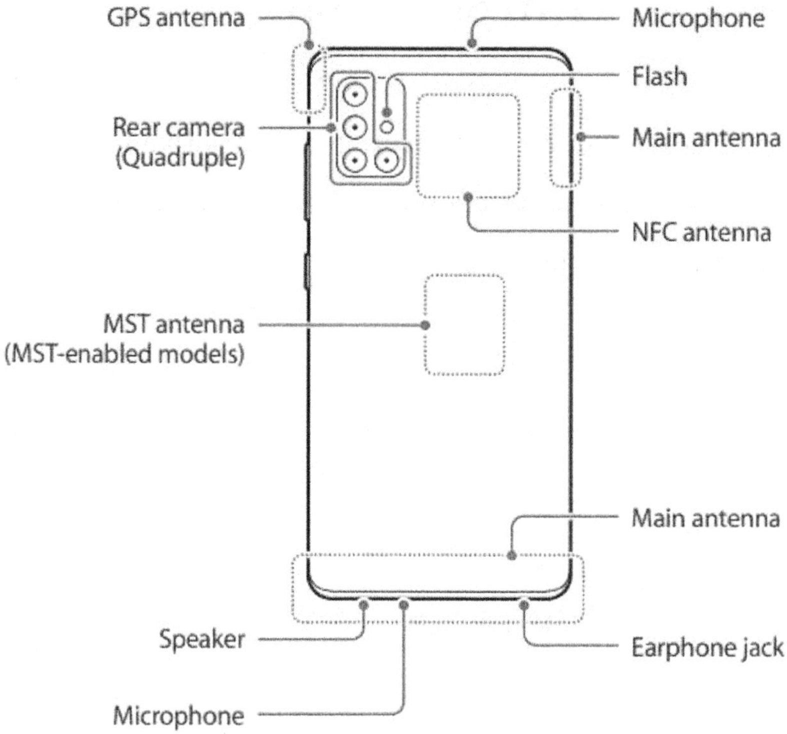

When using the speakers, such as when playing media files or using speakerphone, do not place the device close to your ears.

• Be careful not to expose the camera lens to a strong light source, such as direct sunlight. If the camera lens is exposed to a strong light source, such as direct sunlight, the camera image sensor may be damaged. A damaged image sensor is irreparable and will cause dots or spots in pictures.

• If you use the device with its glass or the acrylic body go broken, there may be a risk of injury. So, ensure you Use the device only after it has been repaired at a Samsung Service Centre.

Connectivity problems and battery drain may occur in the following situations:

– If you attach metallic stickers on the antenna area of the device

– If you attach a device cover made with metallic material to the device.If you cover the device's antenna area with your hands or other objects while using certain features, such as calls or mobile data connection.

• Using a Samsung-approved screen protector is recommended. Unapproved screen protectors may cause the sensors to malfunction. Do not cover the proximity/light sensor area with screen accessories, such as a screen protector, stickers, or a cover. Doing so may cause the sensor to malfunction. The colors on the Edge screen may look different depending on your viewing position.

About Galaxy A71 physical keys

You will find two physical keys on the left side of your Galaxy A71 device. The longer key is the usual Volume key used for adjusting the volume of your device. The shorter key which you will find below is the Side Key with Bixby or power functionality. The side key has different functions which you can access either by long pressing, short pressing or double pressing it.

Depending on the functions, the physical keys can either be used together or alone. Although each key has its own function, however, when pressed together, customized shortcuts can be created. Basic functions of the default key on the left side of the Galaxy A71 device are as follows:

Note: If you don't want to use the side key to switch your device off, you can do that in the Quick Settings panel. Simply find the power icon from the drop down screen at the top.

Click the power icon and tap power off to switch off the phone.

1. Quickly press the side key for the screen to either to switched on or off.
2. To access the Bixby Voice, press and hold the side key.
3. To open the camera, double click on the side key.
4. Press and hold the side key to switch the phone on
5. Press and hold the side key together with the volume down button to switch the phone off.

6. To screenshot, quickly press the side key together with the volume down.

Key	**Function**
Volume key	To adjust the device volume, Press the volume key.
Side key	• Press and hold the side key when the device is off to turn it on. • Press to lock or turn on the screen. • Press and hold to begin a conversation with Bixby. • Press the side key twice or press and hold to open the app or features you set.
Side key Volume + Down key	• Press the side key volume plus down key to simultaneously capture a screenshot. Press and hold at the same time to turn off the device.

Chapter 1: Customize the Side key

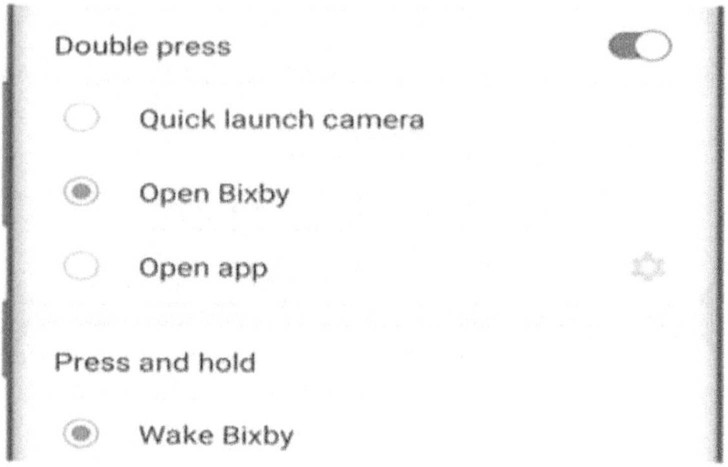

You don't have to get bored with the simple function of your device. The reason is that, it is possible to customize the keys to suit your needs or function the way you want.

How to customize the functions of the side key:

1. Scroll to the Side key option on Settings
2. Click on the side key and initiate your personal configuration.

You can also change the default function for Double press with the following functions:

- Phone switch off
- Open camera

- Open Bixby
- Launch app which will bring out the settings icon to select the app you wish to launch.

You can also change the default settings for Press and hold with two functions: wake up Bixby and open the switch off menu.

Customize the Volume keys with Accessibility or other settings

Getting to the Accessibility option on your Galaxy A71 device is easy if you reconfigure the side key to suit your needs.

Once you've done this, you will be able to gain entry into the Accessibility settings fast.

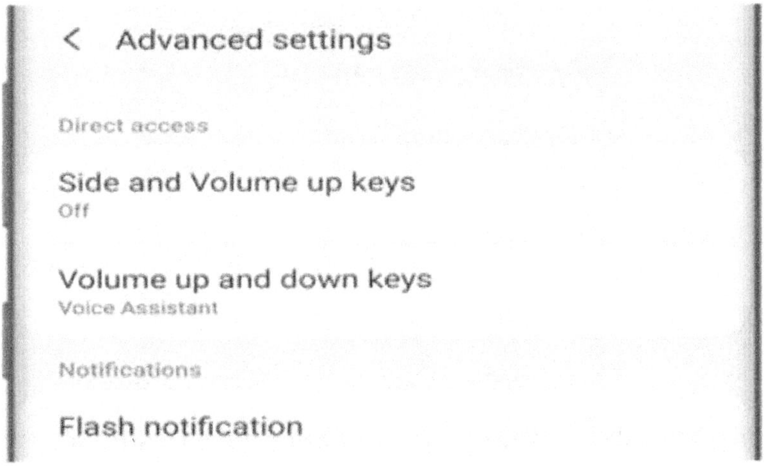

To set this up, go to Settings. From there you will find the advanced settings. Click on it and you will see 'Direct access'.

Under the side and Volume up tab, change the feature default setting by clicking on it and turning it on. Add this new feature to the side key shortcut by pressing the switch beside the Accessibility settings. Try your new shortcut by pressing both the side and volume up keys together. The Accessibility option will pop out and then, you can pick from the option.

Note: You can select just one setting from the Accessibility option to become a shortcut but rather than have a list of settings, you will get just selected setting.

Customize with other features

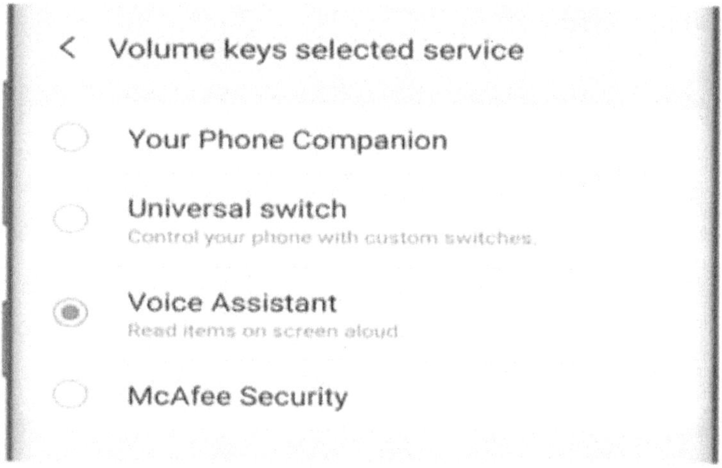

Another physical key that you can reconfigure to suit you style is the Volume up and down keys. You can boost its functionality as you add your most important tools.

Once again, locate the Volume up and down keys in under Direct access by following these steps: Settings > Advanced settings > Direct access > Volume up and down.

You will tap the selected option which will bring out the options:

- Your Phone Companion
- Universal switch
- Voice Assistant
- Security

Change the Button Order Or Grid In The Quick Settings Panel On Your Galaxy Phone

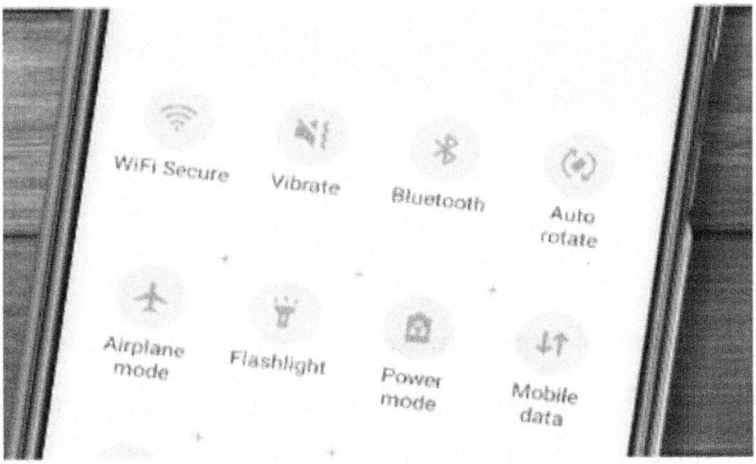

The Galaxy A71 provides you with so many control options but it is not so difficult to sort them properly on the device. You will be able to manage and use these features very fast with ease when you reconfigure it from the Quick Settings board.

To start with, functions like phone vibration and Flight Mode can easily be customized. It is possible to alter the position of the tabs or grid buttons and modify the keys according to your preferred style.

However, you should keep in mind that the type of internet service supplier, OS model, and phone version will determine the type of configurations or screen available.

Change the Button Order

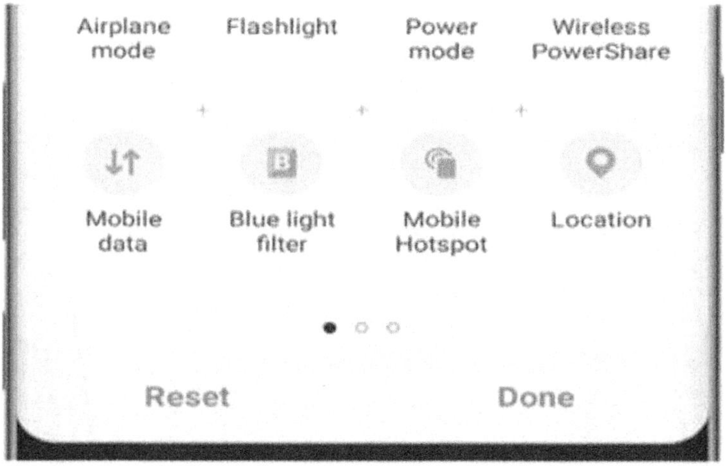

If the Wi-Fi or flashlight of your phone is your major application on your device, simply, alter the position of the key such that these two applications, being your favorite, appear first.

To do this:

- Open the Quick Settings by using your two fingers to drag the screen from the top,

- Click on 'More Options', and then
- Click the 'Button Order'
- Select and hold the button you wish to move and then
- Carry it to the required place.
- Click 'Done'.

Change the Button Grid

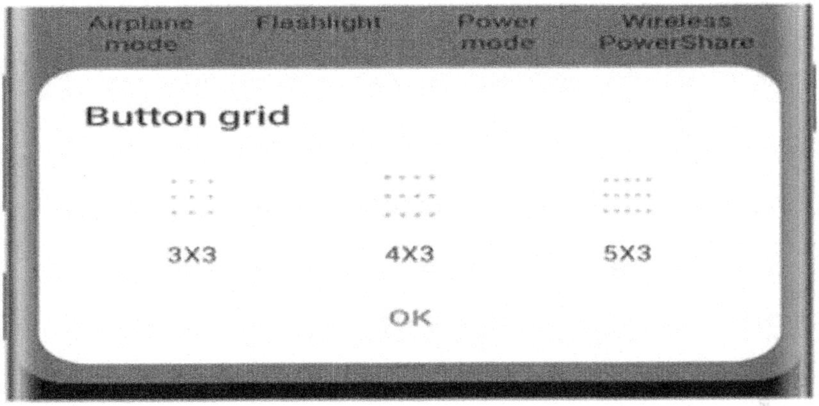

Many users like different buttons to appear in a grid. You will do this by simply altering the structure of the button. It is in fact possible to get up to 15 keys On the Galaxy A71 Quick Settings board. However, you can't get less than nine in a grid. Here is how to go about it.

• Once again, bring the Settings board down by using two fingers to drag from the top to the bottom. Press 'More Options', then

• Press 'Button grid'. And Choose the grid form you prefer from the three options and then

• Click OK.

Chapter 2: Maintain the Battery Life of Your Phone

Your battery is always getting drained off without you even having a clue. Fortunately, the battery life of your phone can be monitored and extended and you can always make use of the Power saving mode feature on your phone as well.

By making use of the above opportunity gifted by the phone, you shouldn't be disappointed about the performance of the battery of your phone again.

Note that based on the wireless provider of your phone, software version as well as your phone model, its available screens and settings do varies.

Battery Usage

To check the amount of battery you must have exhausted over the past week, the battery usage feature enables you to view your estimated battery life.

The feature also enables you to monitor how much battery is exhausted by your apps.

i. Go to Settings.

ii. Swipe to Device care.

iii. Tap on Device care.

iv. Tap on Battery.

v. Go to the top of the page and tap on Battery usage to view your Battery usage.

Screen timeout

There will be a need for your phone timeout sometimes. However, you can control the timing out of your phone to help in saving the battery life.

i. Go to Settings.

ii. Search for Screen timeout and select it.

iii. Select Screen timeout again.

iv. Choose and select your preferred option.

By default, the screen timeout is always set to 30 seconds but it is advisable to set it to 15 seconds to help save its battery life.

Accidental touch protection

Blocking accidental touches will go a long way in reducing battery strain as well as stopping pocket dials that can be quite embarrassing.

i. Go to Settings.

ii. Search for Accidental touch protection and select it.

iii. Tap on the switch which is next to the Accidental touch protection to get it turned on or off.

Adaptive Brightness

Turning on the Adaptive brightness will help in preventing the phone's screen from hurting your eyes by having an understanding of the pattern of your usage. Once you have an understanding of this, you can have the brightness of your screen automatically adjusted depending on your pre-adjustments.

i. Go to the Settings.

ii. Select Display.

iii. Tap on the Adaptive Brightness option.

iv. Tap on its switch to get it turned on or off.

v. Tap on the Reset usage patterns if you need the Adaptive brightness to start all over from the beginning and select Reset for your device to start learning of your usage patterns again.

Power Mode

It is to be noted that the use of the power usage feature can have an effect on your device performance including the app, thereby slowing down the update and completion of some features.

When the power saving mode is activated, note that this might stop background apps from receiving updates as well as sending notifications.

It is advisable to always remember to change the power mode of your phone before going to sleep at night to help in reducing the performance of your phone as well as saving its battery life.

i. Go to Settings.

ii. Swipe to Device care.

iii. Tap on Device care.

iv. Tap on Battery.

v. Tap on the Power mode.

vi. Select your preferred power mode.

vii. Apply your preferred settings or customizations.

viii. Tap Apply.

Here are all the available options for the proper management of power:

1. High performance: This feature maximum screen brightness as well as resolution. It drains more battery life and it should be noted that it's not available on every device.

2. Optimized: This will provide you with the recommended balance of performance as well as battery life.

3. Medium power saving: This helps in extending your battery life by getting some functions limited.

4. Maximum power saving: Helps in saving as much battery life as possible.

Adaptive Wi-Fi

Power too can be saved even if you are always browsing your favorite blogs including video channels on the Wi-Fi network. Also, you can get your Wi-Fi usage analyzed to help in the reduction of your battery usage.

i. Go to Settings.

ii. Select Connections.

iii. Select Wi-Fi.

iv. Select the Advanced option.

v. Tap the switch which is next to Turn on Wi-Fi automatically.

It is however, important to note that this feature isn't available on all phone models.

Getting to fast charge your Galaxy phone

The adaptive fast charger will enable you to get your phone charged up to full power more quickly.

It is the charger that was packed with your phone on purchase which is specifically designed for it to provide you with an extremely fast charging time.

It is however, advisable for you to stick to Samsung approved charging devices as well as batteries. This is because engaging in the usage of other accessories may get your warranty invalidated as well as causing damage to your device.

Turning Fast Cable Charging On

Your phone will automatically turn the fast cable charging on once it detects that a fast charger has been plugged into it. The fast cable charging is automatically turned on by default and you can get it turned on automatically as well.

i. Go to Settings.

ii. Swipe to Device care.

iii. Tap on Device care.

iv. Tap on Battery.

v. Tap on More Options.

vi. Select the Settings option.

vii. Tap on the slider which is next to Fast cable charging.

Fast Cable Charging On

Ensure you are making use of the right charger approved for your phone before starting to charge.

Once you can confirm you have the correct charger, you will need to take the following steps to help your phone get charged at top speed which includes:

1. Get the USB cable inserted into your phone's USB port.
2. Connect the USB cable to the charging adaptor after which you can now get the adaptor plugged into a standard outlet.
3. Get the charger plugged into an outlet to help in enhancing charging time. It is to be noted that getting the USB cable plugged into a computer or another device will only end up prolonging your phone's charging time.
4. Get the charger unplugged from the phone as well as the adaptor from the wall as soon as your phone is fully charged.

You are advised to request service in a situation where the Adaptive Fast Charger doesn't work.

Chapter 3: Inserting Your Sim Card

In the process of inserting your card, it is important to be careful when inserting the ejector tool into the microphone hole which is located at the top of the phone.

To insert a Nano-SIM and the microSD card, you are required to take the following steps:

i. Turn off your phone power if it is on.

ii. Insert the ejector tool that was provided with the phone on purchase into the hole on the SIM/microSD card tray.

iii. Push the tool until the tray pop is out.

iv. Line up your Nano SIM card carefully with the smaller cutout on the tray.

v. Get the microSD card placed on a larger cutout with the same tray.

vi. Make sure the cards are flush with the tray.

vii. Get the SIM/microSD tray inserted back into your phone slot.

viii. Turn your phone power on and you will be set!

It is however, important to note that the microSD cards are supported by all phone models.

Still looking for the power button on your Galaxy A71 device?

It used to be on the right side of the device but finding it should not worry you. It is now located on the left integrated with Bixby functionality.

The Side key as it is now called can be used to switch off your device (Bixby) and carry out other operations based on your customized settings.

Removing the Nano-SIM And Microsd Card

Unmount first before removing the microSD card by taking the following steps:

i. Go to Settings.

ii. Search for Storage and select it.

iii. Tap on the SD card option.

iv. Tap More Options.

v. Tap Storage settings.

vi. Tap Unmount.

Now you can turn off your phone power and detach the Nano-SIM as well as the microSD card by taking the following steps:

i. Get the ejector tool inserted into the hole on the SIM/microSD card tray.

ii. Push until the tray pops out.

iii. Remove the Nano-SIM as well as the microSD card from the tray.

iv. Inset back the SIM/microSD tray into the slot.

SIM card sizes

Note that some carriers might provide you with all three SIM cards. Therefore you must ensure removing the one you need from the pre-scored plastic before getting your phone inserted. You will have a detailed understanding of the SIM card sizes below:

1. **Standard SIM**: This is the largest SIM size and it is also referred to as Mini SIM card more popular in older phones.
2. **Micro SIM**: This is a medium-sized SIM card which is more popular with previous model.
3. **Nano SIM**: This is the smallest SIM size and it's used frequently in most new phones.

Chapter 4: Your Galaxy Phone And Microsd Cards

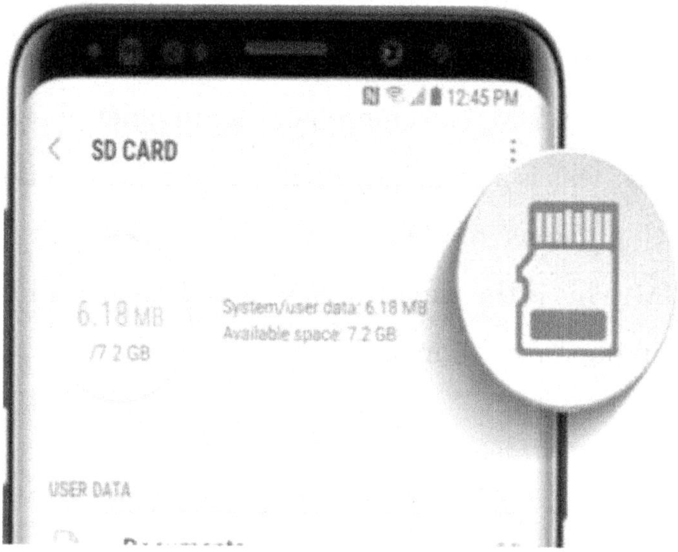

Are you running low on storage on your phone? Having an SDcard on your phone will create more storage room to store more files such as videos, pictures, apps, music and more. You need to have SD cards on your phone to ensure that your pictures and music are stored on it.

Types of MicroSD card

If you are not a tech expert, there is a low chance of knowing that microSD cards have different categories.

I am not talking about the storage capacity. Some other things set their classes apart while some have more unique features that give them a boost in efficiency above others.

Classes of Microsd Cards

MicroSD cards have, at least, up to 2 GB storage capacity. microSDHC (High Capacity) cards can boast of a storage capacity that is up to 32 GB. While microSDXC (extended Capacity) cards can be over 32 GB to 2 TB storage capacities. Speed rating among each card is 2, 4, 6, 8, or 10 ranging from the slowest to the fastest.

Even though it is possible to use any type of SD card on your phone, the most efficient for your phone is the **microSDXC** because it gives you high-resolution video recording and that is because of its large storage capacity and its great reading speed. If you are not using your phone for heavy task like videoing, the less efficient card such as microSD or **microSDHC** should be enough.

Inserting MicroSD card

MicroSD cards offer an additional storage on your phone. But they are very small and tiny, therefore, the need to be careful not to lose your microSD card. These two easy steps will teach you how you can insert your SD card into your phone.

Note: All models of phones supports micro SD cards.

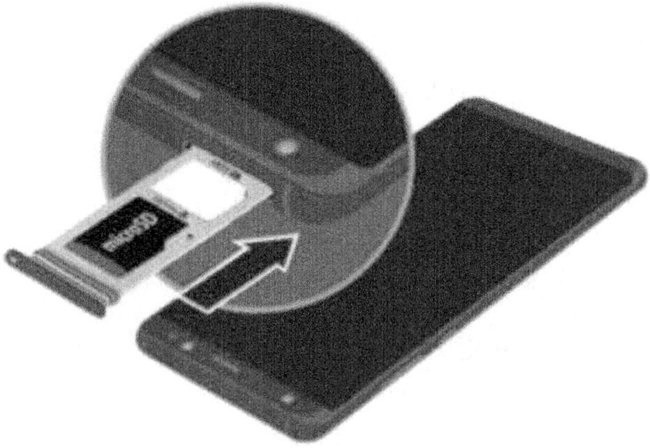

There is an ejector tool that comes with your phone, carefully insert this tool into the SIM and SD card tray to slide it out of the phone.

Note: Do not mistake the SIM hole to the microphone hole because they are both located on top of your phone.

Gently place your SD card on the tray and slide it back in.

Note: The slot for the SIM is the smaller opening below.

Transferring Media to A Microsd Card

You do not need a moving crew; you can transfer your files to a microSD card. There is an option in the **Device Care** that allows you to free some space.

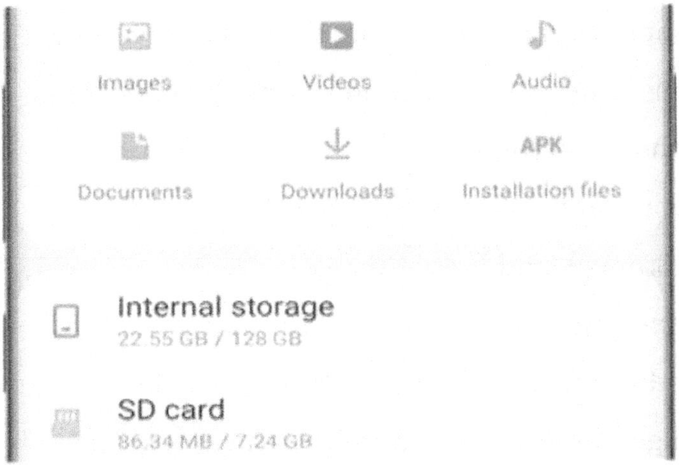

1. The first step is to insert your microSD card, and stroll to the Samsung folder and click My Files.
2. Look for the files you would like to move and make a long hold on the file(s).
3. Click the Move or Copy below the screen and go back to the **'My Files' home**.
4. Click on the SD card this time and go to the folder you would like to move the files.

Note: This method is only possible when you are moving files using the My Files app. It is impossible to move files to other apps from My Files.

Moving applications to microSD card

Before moving an app to a new location, ensure that there is SD card inserted. Or else, the moving action will not be possible.

1. Go to Settings, and click Apps.
2. Go to the app you want to move and select it.
3. After this, click Storage and press Change.
4. Select SD card and press Move to switch the app's location.

If you cannot find the Change option available, it will be impossible to change the storage location. For instance, Apps that are preloaded can't be moved.

Removing the Microsd Card

At some point in time, you will need to remove the microSD card from your phone, maybe you want to change to an upgraded one, whatever reason you are removing your SD card, here is how to go about it to ensure that you are doing it correctly.

Note: Remember to unmount the MicroSD card before you eject it.

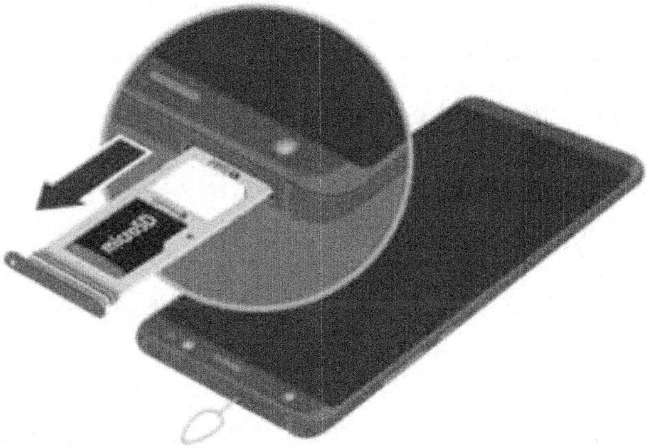

1 Go to Apps and click on Settings.

2. Click Search, and look for Storage, click on it when you have found it.

3. Click More Options and then click Storage settings.

4. Click Unmount and carefully insert the ejector tool into the SIM and microSD hole tray on the top of your phone.

5. When the tray comes out, pull out the microSD.

Note: Do not mistake the SIM hole to the microphone hole because they are both located on top of your phone.

Encrypting and Decrypting an SD Card

In a situation where you have data to hide, you can go into the spy mode and have the SD card on your phone encrypted.

i.Go to Settings.

ii. Tap Biometrics and security.

iii. Tap on the Encrypt SD card option.

iv. In case the screen lock is set up on your phone, your credentials may need to be entered.

v. Review information about getting a microSD card encrypted and select Encrypt SD card.

vi. If you would like to decrypt the card instead, select Decrypt SD card.

It is to be noted that the setting may not come up unless you have inserted the SD card into your phone.

Performing Factory Reset on Your Galaxy Phone

In case any reason arises for you get your phone data wiped, a factory reset can be performed. This feature will enable you to completely get your personal information and data erased which will keep your phone in a clean state.

Resetting the device

The first thing to do before factory resetting your phone is getting all your important information saved to prevent you from losing them.

In a case where you have gotten your microSD card encrypted, it must be decrypted before resetting your phone, otherwise, the data on the card will be unreadable. To do this;

i. Go to Settings.

ii. Search for Decrypt SD card and select it.

iii. Follow the on-screen instruction prompts afterward.

Ensure backing up your data to prevent you from losing them before performing a factory reset. To perform a factory reset, you are required to take the following steps:

i. Go to the Settings.

ii. Search for the Factory data reset and select it.

iii. Tap on the Factory data reset again.

iv. Review the information.

v. When done, swipe up to the Reset and tap on it.

vi. In a situation where you have set up a security lock on your phone, you will be prompted to enter your credentials.

vii. Tap on Delete all to get all information and downloaded apps deleted from your phone.

Once you have followed these steps accurately, your phone will reset when turned back on by starting from the initial setup screen.

Chapter 5: Turning the A Series On And Off

Use the Power key to turn your device on or off. Do not use the device if the body is cracked or broken. Use the device only after it has been repaired.

1. Press and hold the power button to power on the device.

2. Follow the on-screen instruction on your device to set it up upon booting up the device at first or after a data reset.

3. Press and hold the power button for a few seconds and click in the Power off option to shut down the device.

4. Adhere to every precautions and guidelines from recommended personnel in wireless devices restricted areas, like gas stations, airplane and hospitals.

Use the Setup Wizard

When you turn on your device for the first time, the Setup Wizard shows you the basics of setting up your phone. Adhere to the prompts to:

- Select a default language
- Then to connect to a Wi-Fi® network,

- Set up accounts,
- Select location services,
- Learn about your device's features, and more.

Restarting the Device

1. Tap and hold the power key and select the Restart option.

2. In cases whereby you experience unresponsiveness and freezing, Tap and hold the power and the Volume Down button simultaneously for about 7 seconds to automatically restart it.

Emergency Mode

In the case whereby you are minimizing consumption of battery, you can switch your device on to emergency mode. However, there are probabilities that some functions and apps may not function, or function minimally while in emergency mode. You are only liable to make emergency calls, send location details to others, sound emergency alarms as well as many of such vices while in the emergency mode.

Activating Emergency Mode

For you to activate the emergency mode, Tap and hold the Power key, and then select the Emergency mode option. You'll see the time range left before the battery power finally shuts down. This of course varies in different device depending on the operating and setting conditions.

Deactivating Emergency Mode

If you want to deactivate the emergency mode, click on ⋮→ **Turn off Emergency mode**. More conversely, Tap and hold the Power button, and then deselect the Emergency mode.

Initial Setup

Follow the on-screen instruction on your device to set it up upon booting up the device at first or after a data reset.

The initial setup process varies depending on your location and installed software.

1. Switch the device on. Choose your preferred language and click on the option to continue.

1. Go through the terms and conditions about the device and operation and click on Next.

2. Transfer your content from a previous device and move to your device.

3. Connect to functioning Wi-Fi network connection

4. Inability to connect to a working Wi-Fi network would limit your accessibility to some features while setting up your device initially.

5. Adhere to the on-screen guidelines to complete your initial setup. For device protection, select a preferred screen lock method to secure your device. This way, you are preventing third party intrusion to your private files on your device. However, if you are not comfortable setting up your device

protection immediately, you can save it up for later. Click on the Not now option. If you get a notification of recommended apps, you can choose the option of the ones you want and discard the ones you don't want or in more crucial cases, avoid selecting anyone at all and move on with your setup process.

7. Sign in to your device's manufacturer account. This way, you can enjoy every one of their services; update your data as well as securing every part of your device. Go to the manufacturer's website to access more information.

You can add and sync all your email, social networking, and picture and video sharing accounts. From Settings, tap Accounts and backup > Accounts > Add account. Tap one of the account types. Follow the prompts to enter your credentials and set up the account. Tap Auto sync data to enable automatic updates to your accounts.

Chapter 6: Samsung Account

Your Samsung account is an integrated account service that allows you to use a variety of Samsung services provided by mobile devices, TVs, and the Samsung website. Once your Samsung account has been registered, you can keep your data up to date and secure across your Samsung devices.

For more information on Samsung accounts, launch the **Settings** app and tap **Accounts and backup** → **Accounts** → **Samsung account** →→ Help.

Registering New Samsung Account

If you do not have a Samsung account, ensure you create one for yourself.

1. Open the Settings app and tap Accounts and backup → Accounts → Add account → Samsung account.

Alternatively, launch the **Settings** app and tap

2. Tap Create account.

3. Follow the on-screen instructions to finish creating your account.

Registering an Existing Samsung Account

If you already owned a Samsung account, register it to the device.

1. Open the Settings app and tap Accounts and backup → Accounts → Add account → Samsung account.

Alternatively, launch the **Settings** app and tap

2. Fill in your Samsung account ID and password and tap Sign in.

2. Anytime you fail to recall your account information, Reset password or tap **Find ID** to find the account information.

3. Read, then agree to the terms and condition and click on next to end your Samsung account registration.

How to Remove Your Samsung Account

Note that when you delete your registered Samsung account from the device, most of your sensitive data, like events, contacts will also be deleted. To do this:

1. Open the **Settings** app and tap **Accounts and backup → Accounts**.

2. Tap Samsung account → ⋮ → Remove account.

3. Tap Remove, then fill in your Samsung account password, and then tap OK.

Samsung Members

Samsung Members provide support services to customers, such as device problem diagnosis, and allows users to submit questions and error reports.

You can also view the latest Galaxy news and tips or share information with other customers in the Galaxy users' community. Samsung Members can provide help to solve any problems you might encounter while using your device.

- This app's availability and supported features are variable and depend on the region or service provider.

- To post your comments or submit your feedback, you must register and sign in to your Samsung account.

Chapter 7: How to Transfer with Samsung Smart Switch

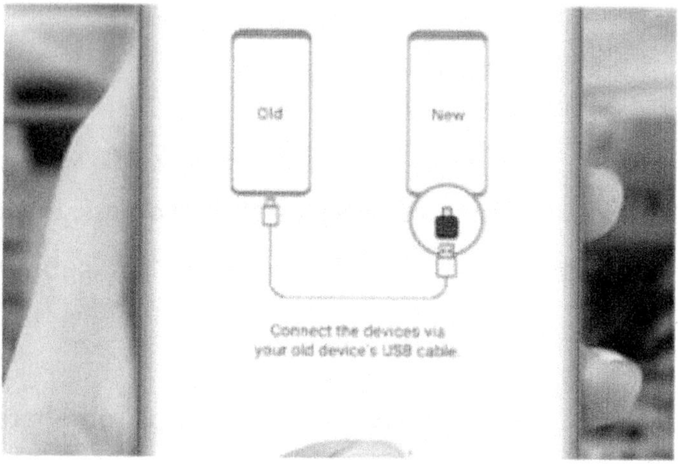

Nothing beats the feeling when you just got yourself a new phone except the ability to get your files from your old phone back into the new one. That is exactly what Samsung Smart Switch does, it lets you retrieve your files from your former to the current be it photos, videos, and even your contacts.

With Samsung Smart Switch, you can transfer between various smart phones such as Android, Blackberry, iOS and windows.

Transferring Content Using Wi-Fi

Before anything, make sure the Smart Switch app is installed on both phones. When you're ready, place the phones within 4 inches to each other.

1. Open the Smart Switch on the phones respectively. Click on the send data on your old phone and press Wireless. This action will make both phones connect automatically.

2. Select the files you want to share with the new phone and press Send.

3. Now click receive on the new phone, this will transfer the files selected to your new phone.

4. When the transfer is completed, tap done on the new phone and press Close on the old phone.

5. Click to launch the Smart Switch app on your old device.

6. Press the SD card icon visible in the top right corner and press Backup.

7. Choose those files you would like to back up and click on the Backup soft button again.

8. After the backup is successful click Done.

The step to take next is to connect the external storage device to your new phone and open Smart Switch.

1. Press the SD card icon and click on Restore.

2. Choose the files you want to restore and press **Restore** after you have made your selection.

3. After restoration is successful, click **Done**.

Transferring Content Using Icloud

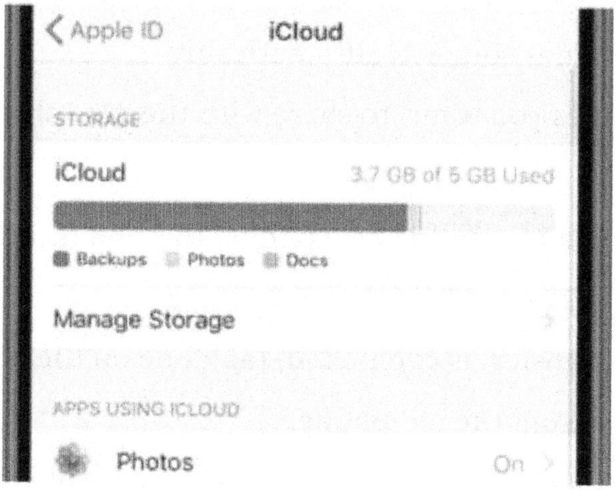

There are important files stored on your former iOS device and moving it to your new phone seems to be a daunting task.

Do not fret; Smart Switch has made it easy and possible for you to carry out your transfer action with iCloud.

1. Open the Smart Switch on your new device and press the Receive soft key.
2. Click on Wireless, after that, you should press iPad or phone.
3. Carefully input your Apple login ID to sign in. Input the verification code correctly and press OK.
4. Choose the files you want to transfer and press Import and press Import again.

Smart Switch Transfer's Cloud Sign in Screen

Note: It is impossible to transfer iTunes videos and music from the cloud. You should put Smart Switch to use for PC or Mac if you won't transfer iTunes videos or music.

Content Transfer from Windows and Blackberry Phones

Are you switching from Windows or a Blackberry phone to a Galaxy phone? No worries, it is possible to transfer you old files using Smart Switch.

1. Make sure Smart Switch is downloaded and installed on your Windows or Blackberry device.
2. Then open Smart Switch on both devices and click on Receive data on your new device.
3. Press Wireless and press Blackberry or Windows phone.

4. Now on the old phone, click Connect and choose the preferred network. Input the password that appears on your new device. This will transfer your files.

Note: You will need to activate the Mobile Hotspot for wireless connection.

Smart Switch Settings Management

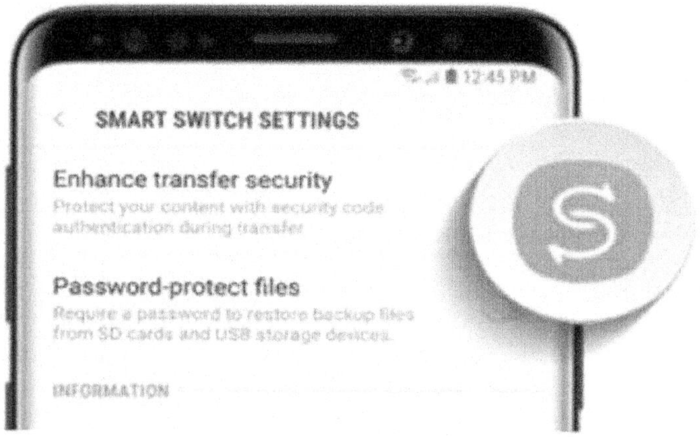

What else is left to know about the Smart Switch app? There are still more to know, one of which is the option available to protect your backed up files during the transfer. Interesting right? Yes it is.

Smart Switch Settings

Ensure your data are kept safe during the transfer process. Find out the Smart Switch settings available for you:

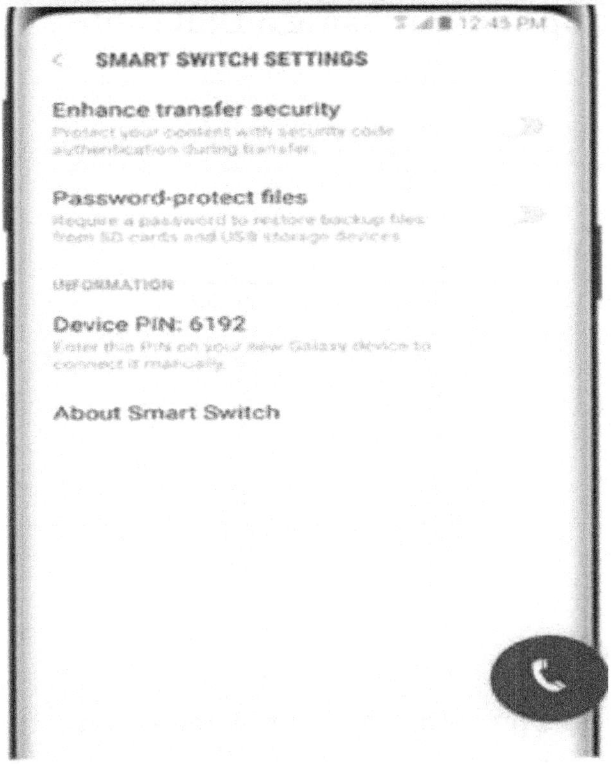

1. Click to launch Smart Switch and then go to **More Options**, and then access the Settings.

The settings available are as follows:

Enhance transfer security: This option secures your content with security code authentication while the transfer is going on.

Password-protect files: This option will ask for a pin before you can restore files from SD card or USB storage devices.

Transferring content using a USB cable

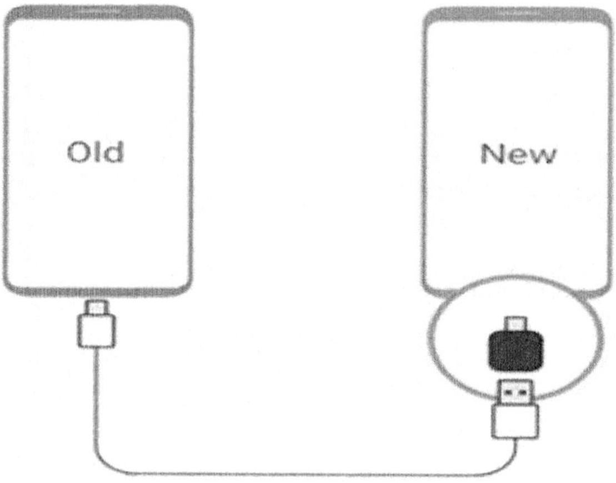

Connect the devices via your old device's USB cable.

In case there is a problem with your wireless connection, the alternate way is using a USB cable. The first thing you want to do is to connect the USB cable of the old phone with the new phone's USB-OTG adapter.

1. Open Smart Switch on the phones, then click Send on your old phone, and click Cable.
2. Press Receive data on your new device. After it is done scanning your old device, make a selection of the data you want to share to your new device and press Transfer.
3. On the old device, click on the Done soft button and close.
4. Note: You are advised to use Wi-Fi to transfer large files and ensure that both devices are well charged before the transfer.

Transferring Content Using an External Storage

If you have some important files stored on your microSD card, you can transfer them using Smart Switch. But before the commencement, make sure the microSD or USB-OTD is inserted into the old device.

Device PIN: Input the PIN on your new Galaxy device to manually connect.

About Smart Switch: This option lets you know more about the Smart Switch app.

Enhance transfer security: This one is to secure your files with authentication code which you must input during the transfer process.

Password-protect files: A password must be entered when restoring backup files from SD cards and USB storage devices.

Device PIN: Input this PIN if you are connecting your new Galaxy phone manually.

About Smart Switch: This exhibits the app version.

Using the Direct Share on Your Galaxy Phone

The Direct share feature enables you to access and share your photos, videos, etc, instantly by making use of the app's sharing panel and sending contents to the specific contacts. It provides a fast and easier way of getting information shared

Turn on Direct share

With the Direct share turned on, you can easily share your favorite GIF with everyone.

It is to be noted that the Direct share is turned on by default. To get it activated manually, you are required to take the following steps:

i. Go to Settings.

ii. Search for Direct share and select it.

iii. Tap on the switch to turn on Direct share.

Sharing with Direct share

i. Go to the file you like to share for instance open Gallery.

ii. Select and hold the image you like to share.

iii. Select the Share icon.

iv. Go to the sharing panel and tap on the desired option.

It is to be noted that contacts you have shared content with before will automatically appear as icons in the sharing panel.

Chapter 8: Understanding the Screen

Here are some tips to keep in mind when using the Samsung Galaxy A Series screen in order to get the best experience.

- Avoid using the touchscreen to get in contact with other electrical devices. The reason is that Electrostatic discharges can damage the touchscreen and cause it to malfunction.

- Also, avoid tapping the touchscreen with anything sharp as this may damage the touchscreen.

- You also don't want to apply excessive pressure to the touchscreen with your fingertips.

- Do not use fixed graphics on any part or even all of the touchscreen for a long period of time. Otherwise, this may lead to after images (screen burn-in) or ghosting. The device may not recognize touch inputs close to the edges of the screen, which are outside of the touch input area.

- It is recommended to use fingers when you use the touchscreen.

Tapping

Tap the screen.

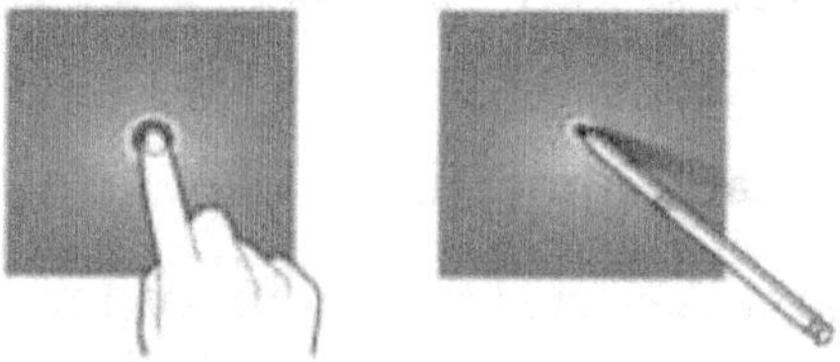

Tapping and holding

Tap and hold the screen for approximately 2 seconds.

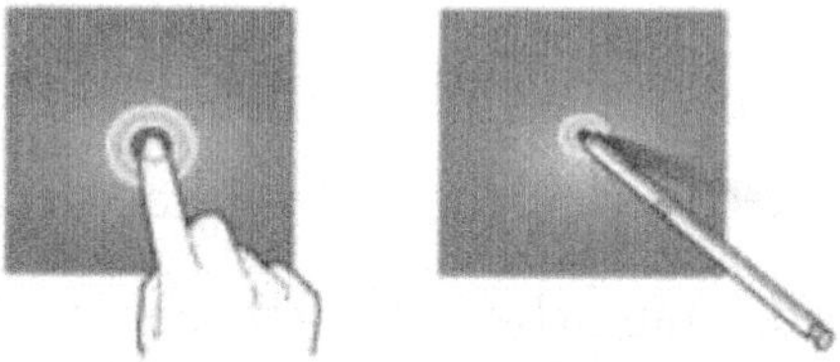

Dragging

Tap and hold an item and drag it to the target position.

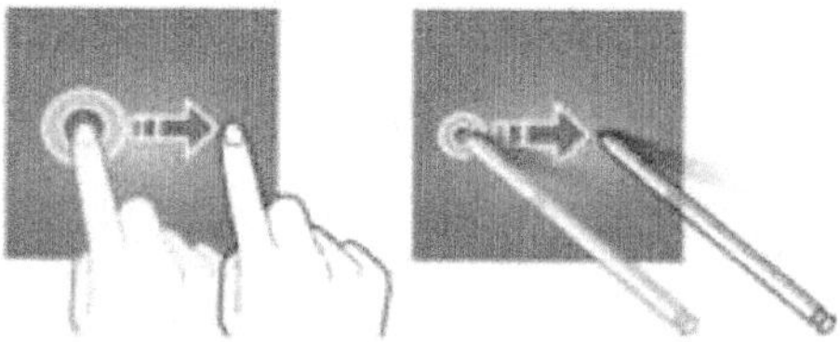

Double-tapping

Double-tap the screen.

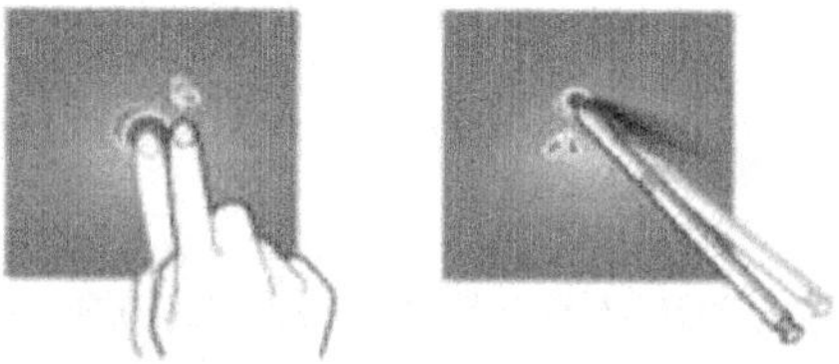

Swiping

Swipe upwards, downwards, to the left, or to the right.

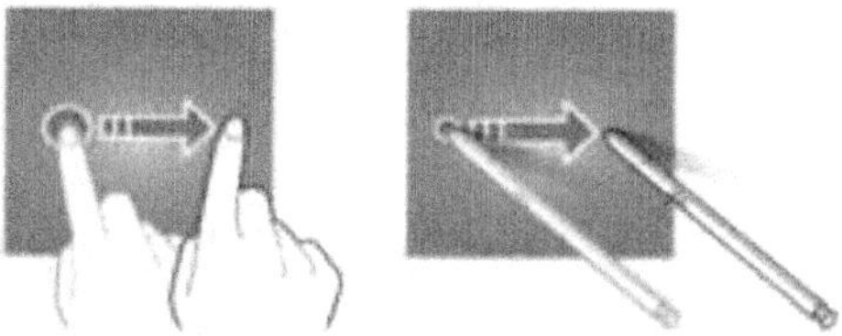

Spreading and pinching

Spread two fingers apart or pinch on the screen.

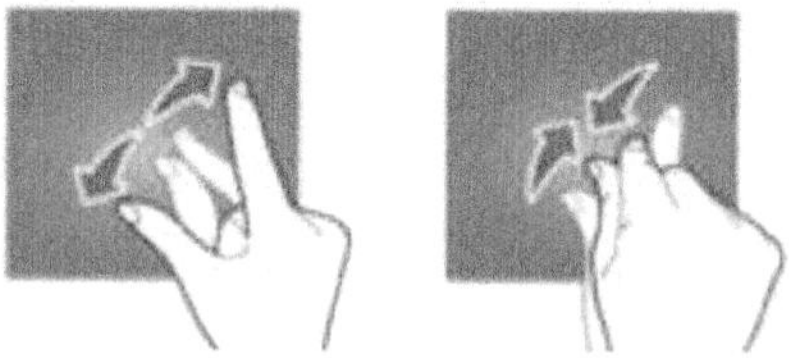

How to Use the Navigation Bar **(Soft Buttons)**

Anytime you turn on the screen, soft buttons will show up on the navigation bar located at the bottom of the screen. The soft buttons are positioned to the Home button, Recent buttons and Back button by default.

Note that the way the buttons work can change based on the app you are currently using or according to the environment.

Soft buttons

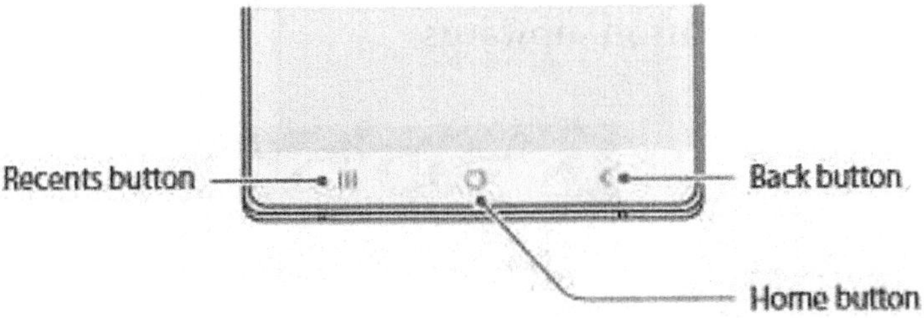

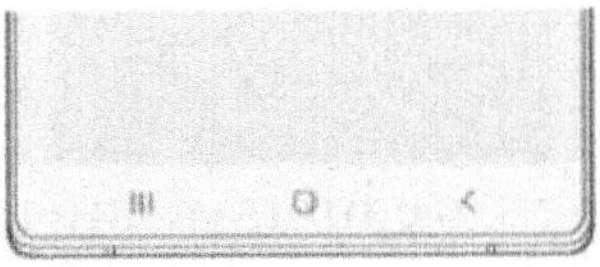

Button		Function
|||	Recents	• Tap to open the list of recent apps.
O	Home	• Tap to return to the Home screen. • Tap and hold to launch the **Google Assistant** app.
<	Back	• Tap to return to the previous screen.

How to Hide the Navigation Bar

You can hide the navigation bar and still view files or use apps conveniently on a wider screen. To do this:

1. Open the **Settings** app, and tap **Display** → **Navigation bar**.

2. From Navigation bar, tap **Full screen gestures** that are under **Navigation type**. This will hide the navigation bar and will show up the gesture hints, which houses the soft buttons.

You can use the soft buttons, by dragging the gesture hint of your desired button upwards.

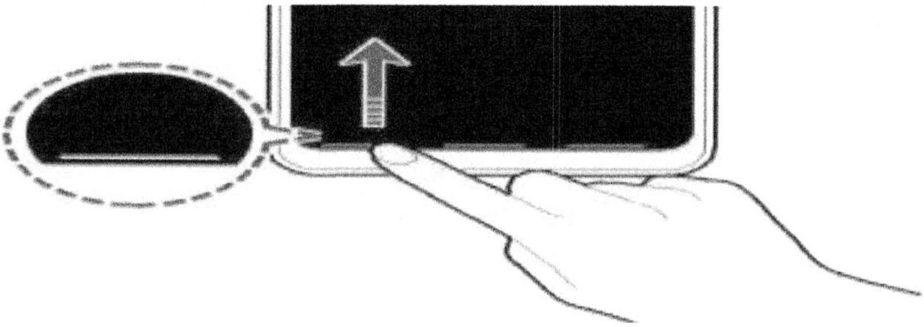

You can also hide the gesture hints at the bottom of the screen, by tapping the Gesture hints switch as this will deactivate it.

Setting the Navigation Bar

To set the Navigation bar to your desired preference;

1. Launch the Settings app,

2. Then tap Display → Navigation bar

3. from there, select an option.

The options you will see below will range from:

• **Navigation type**: This setting allows you to hide or display the navigation bar.

If you intend to hide the navigation bar, it is pretty straightforward:

a. Just Tap Full screen gestures.

b. Once you have hidden the navigation bar, then drag the gesture hint of your desired button upwards to use the soft button.

• **Button order:** The button order setting allows you to change the order of the buttons on the navigation bar.

• **Gesture hints:** This lets you program the device to display indicators at the bottom of the screen to show the locations of the soft buttons. You can only see this feature when you choose to use Full screen gestures.

Customizing the Navigation Bar

Go to Settings app, click on Display → Navigation bar and choose an option.

• **Navigation type:** You can customize your device to hide or show the navigation bar. Click on the Full screen gestures to hide the navigation bar.

•**Button order:** You can use this feature to customize the order at which you want your buttons to be.

•**Gesture hints:** This allows you to display the indicators at the bottom of the screen to appear.

My Files Options

Use My Files options to search, edit, clear file history, and more.

○ From My Files, the following options are available:

• **Search**: Search for a file or folder.

• More options:

– **Clear recent files:** Remove the list of recently accessed files. This option is only available after a file has been opened through **My Files.**

– Storage analysis: See what's taking up space in your storage.

Chapter 9: Setting up the Ultrasonic fingerprint sensor on your Galaxy phone for use

The biometric is a well-thought technology created for advanced security. You can be rest assured that once your Samsung Galaxy is locked with fingerprint security, it is well secured. Nonetheless, the fingerprint scanner is bound to vary depending on the type of your phone.

Note: Your wireless service provider, phone model or software version will determine your screen and settings variety. To be able to add your fingerprint, you must have a lock screen set up.

Set Up and Use Fingerprint Security

Note: The fingerprint unlock feature will be activated automatically when you add a fingerprint.

Set up and use fingerprint security on Galaxy A Series

Note: The fingerprint unlock feature will be activated automatically when you add a fingerprint.

To set it up, follow the instructions below.

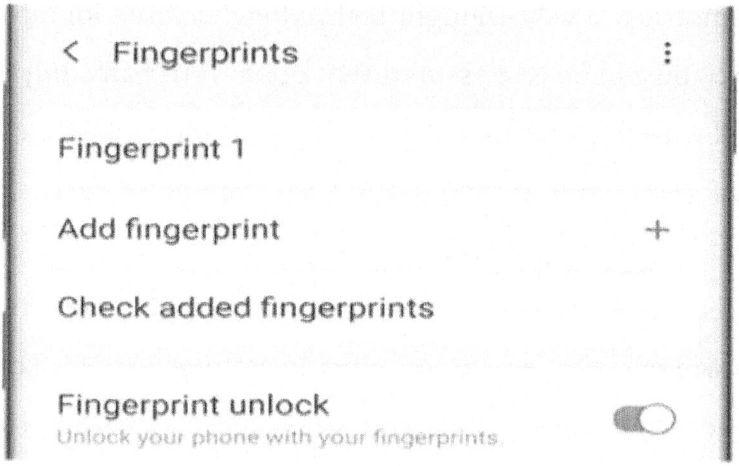

In the Settings, look for Fingerprints and click on it, then input your security details. The adhere to the instruction that appears on the screen to add your fingerprint. When you are done, click on the Add and add another. Click on 'Done' when the process is completed.

If you want to unlock your device, click on the screen or Power button and register with the finger you used for the security set-up. This will let the phone unlock automatically once it has recognized your print.

Note: There is a preinstalled plastic screen protector that comes with the GalaxyA Series. That is why you are advised not to install any screen protector as it could affect the fingerprint sensor.

Removing a Fingerprint From Your Phone

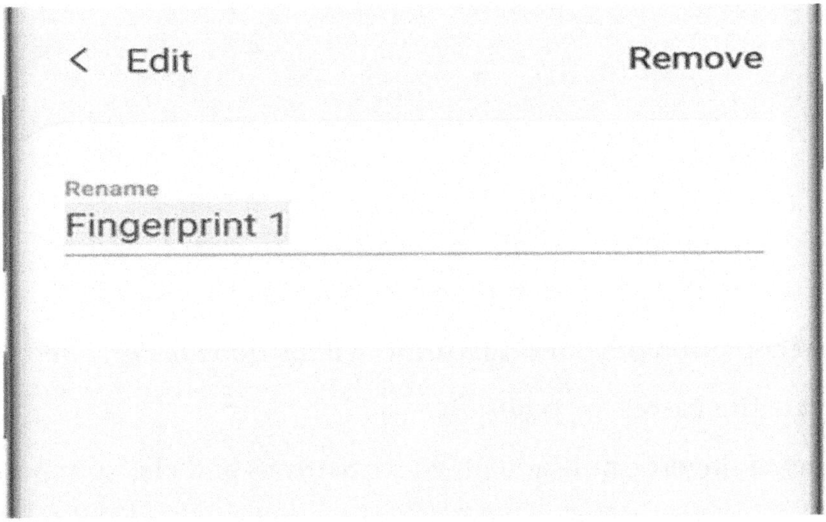

Even though the fingerprint sensors are located in different places, the process of removing the fingerprint is no different with the Pie update on all devices.

In the Settings, look for Fingerprints and click on it, then input your security details. Click the fingerprint you would like to delete for example, Fingerprint 1. Then click Remove. After that click Remove again for confirmation.

Note: The fingerprint security feature will be deactivated when you remove the registered fingerprint.

Turning on Faster recognition on Galaxy Phone

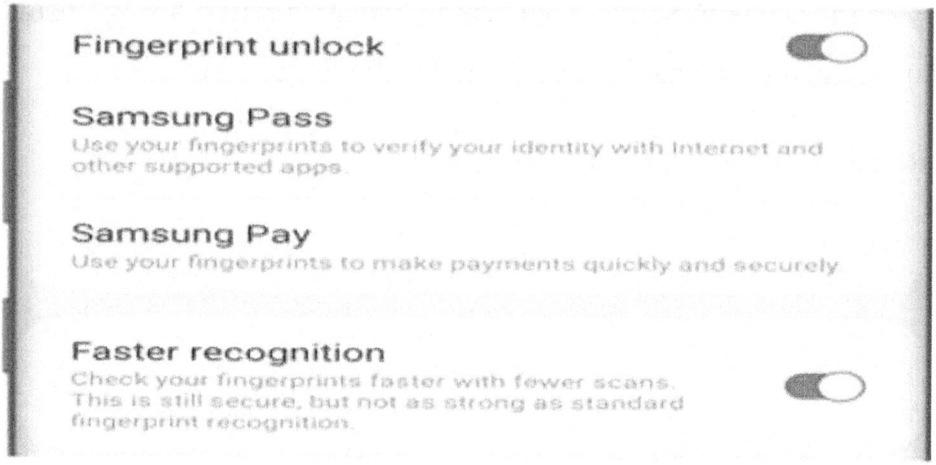

Are you interested in extending the speed of your Galaxy A71? The recognition of your fingerprint will be done faster when you activate the Faster recognition.

1 When a fingerprint is added, go to Settings and click on Biometrics and security soft key.

2. Scroll to fingerprint and click on it and input your lock screen details.

3. Below the screen, you will find the Faster recognition, click on it to get it activated or vice versa.

How to Alter the Secure Lock Settings on Your Device

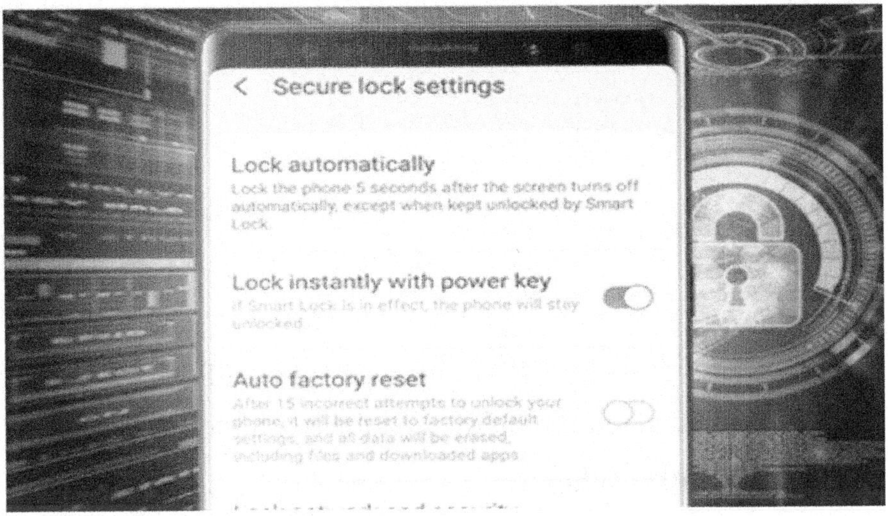

The Secure Lock settings features give you an unwavering security system that put your mind at rest that no one will ever be able to access your phone. It locks the screen automatically according to your preset.

On Auto Factory Reset, there are more means for you to secure your device.

Note: Your wireless service provider, phone model or software version will determine your screen and settings variety.

Secure Lock Settings and Features

Important: The secure lock screen can only be available when the lock has been set.

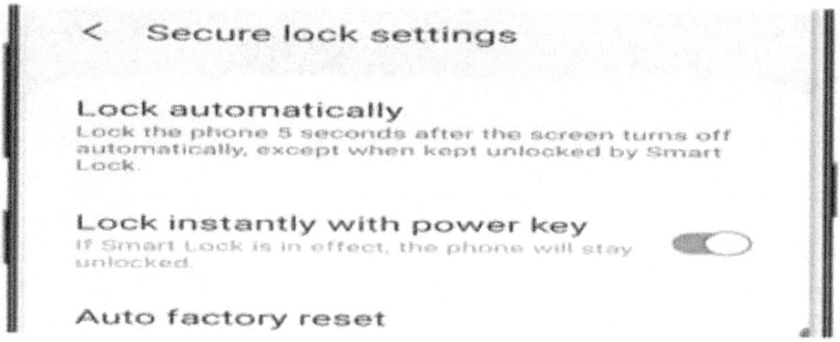

Go to Settings and look for Secure lock settings. Input your secure lock details. Then the below will be available.

Lock automatically: When the screen backlight is off, the screen will lock automatically after a period of a few seconds.

Lock instantly with power Key: The screen will manually be locked when you push the power button.

Note: The phone will be on lock when the Smart Lock is activated.

Auto factory reset: Your phone will reset automatically to factory settings after 15 failed trials to unlock your phone.

Lock network and security: When you phone is on lock, all the security functions and networks will not be accessible.

That is to say, you cannot have control over the mobile data and Wi-Fi, which makes it easy to protect your phone even when lost.

Show Lockdown option: This will show the Power key option that disables Smart Lock, the notifications on the Lock screen and biometrics unlock.

– **Settings:** View settings for the app.

Add or Remove Apps and Widgets on Your Galaxy Phone's Home Screen

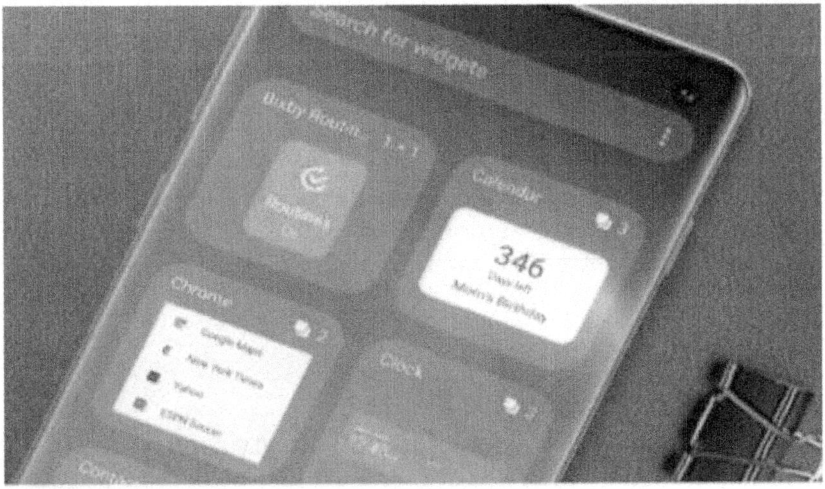

Apps and widgets: is there any difference? You open an app when you press the app icon but a widget is a shortcut for an app which is only active on the home screen.

Add or Remove Apps on a Home Screen

It is not easy to locate an app when there are so many of them installed on your phone. If you want to locate important apps easily, it is best to place them on the Home page.

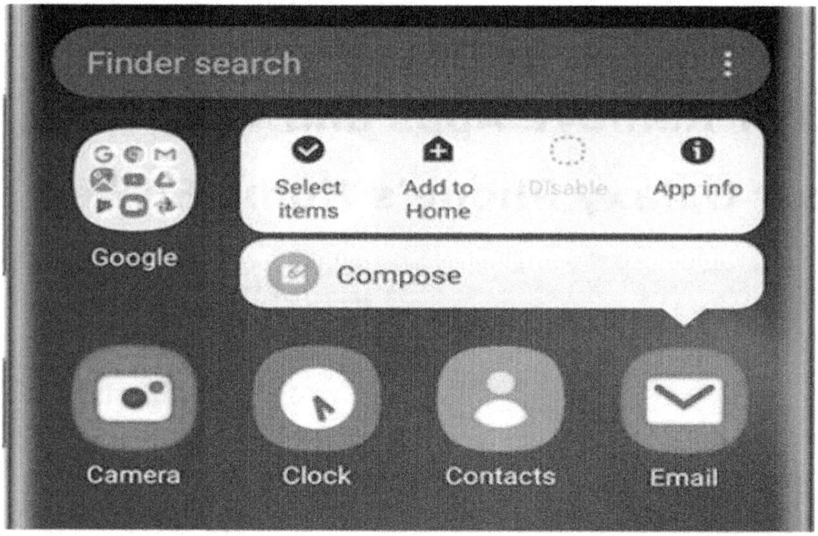

To do this:

- Open the Apps screen by swiping an empty area from the home screen.

- Press and hold the app you want.

- As you hold, carry the app home screen, placing it where you want it to be. You can also do this by long pressing the app and from the options, select '**Add to Home**'. If you find out that there are some apps on your home screen which you no longer use or need, you can remove them. Simply long press the app and press '**Remove from Home**' to remove the app.

Add widgets on the Home screen

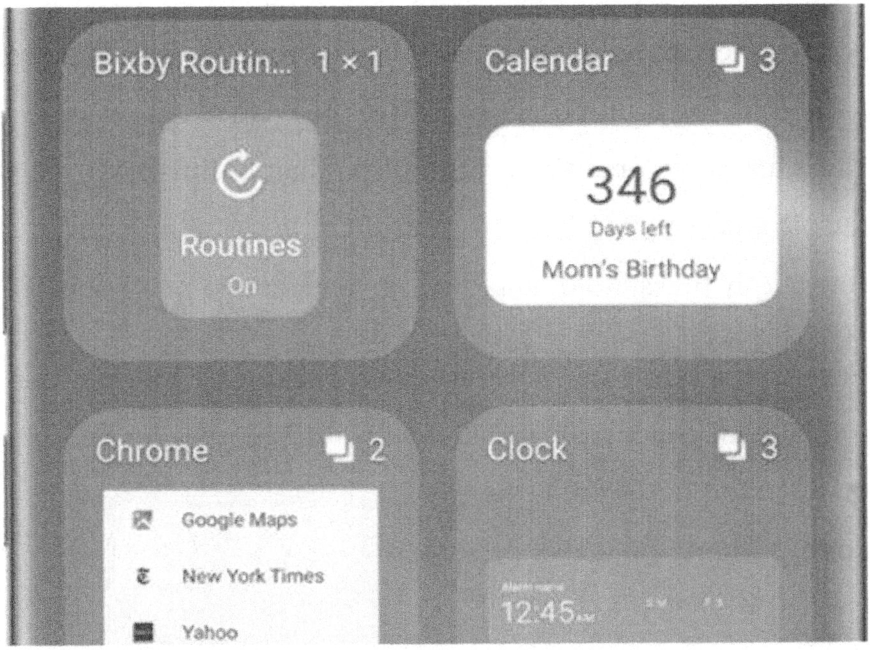

What is a Widget? This is feature that helps you to view all the apps on your Home screen. With a widget, you can gain information about an app without opening it. To place a widget on your home screen:

- Press 'Widgets' from the options you see when you've long press an empty area on the home screen.

- Press and hold the widget you want and carry it to your preferred place on the home screen.

Not all widgets have the same features. While some have different versions, others need some reconfigurations. There is always a frame around a widget that can be resized. Simply press, hold and drag the frame edges to resize the widget. Afterwards, you press an empty area away from the widget to finish the process. In case you want a new position or location for your widget, simply press, hold and drag the widget to a new position.

Note: provided that your home screen is filled with widgets, the next widget you place will automatically find another home screen to stay.

Remove Widgets on a Home Screen

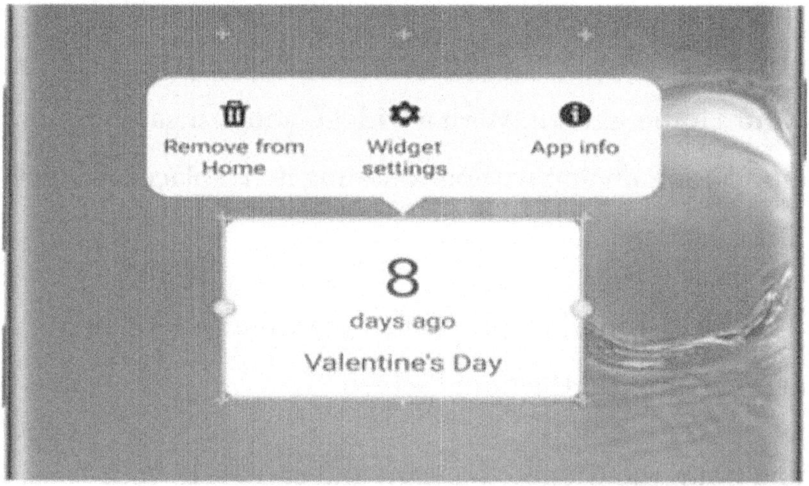

Are you bothered about the number of widgets on your home screen?

Although, Widgets are important however when they are too much on your Home screen, it can be a little nasty. To remove some old unnecessary widgets, simply long press the unwanted widget on your home screen and press 'Remove'

Creating Folders

You can easily create folders in your Samsung Galaxy A SERIES and combine similar apps to let you instantly access and launch apps.

To do this:

1. On the Apps screen, or Home screen, tap and hold an app, and then drag that app over another app.

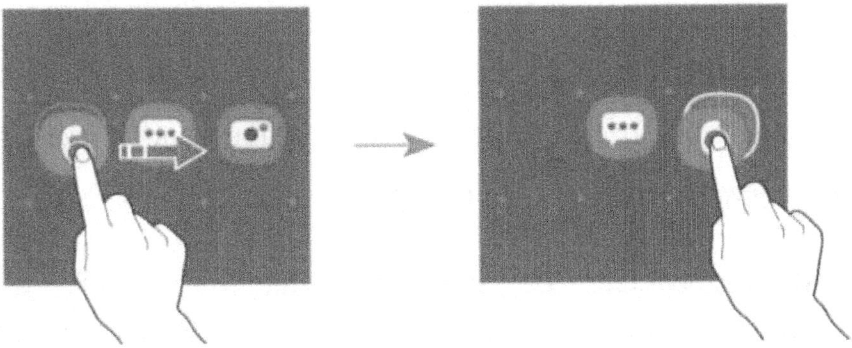

With that, you have created a new folder, containing the selected apps. Then, Tap **Enter folder name and** enter your folder name.

• **Adding more apps**: To add more apps, Tap Add apps on the folder. Then, Tick the apps to add and tap Add. Drag an App to add it to the folder.

• **Moving apps from a folder**: To move apps from a folder to another one, simply tap and hold the app and drag it to a new folder.

• **Deleting a folder:** Delete a folder by tapping and holding the folder, and then tap **Delete folder**. This will have the folder deleted, while the folder's apps will move to the Apps screen.

Editing the Home Screen

To edit the Home screen, you have to pinch your fingers together or tap and hold an empty area to access the editing options.

This allows you to add widgets, set wallpaper and more. You can also delete, add or rearrange Home screen panels.

• **Adding panels**: You need to Swipe to the left, and then tap ⊕.

• **Moving panels**: You will have to tap and hold a panel preview, and then drag it to a new location.

• **Deleting panels:** Tap 🗑 on the panel.

• **Widgets:** Widgets are known to be small apps that launch exact app functions to give information and suitable access on your Home screen. You can add a widget on the Home Screen when you tap and hold a widget, and then drag it to the Home screen.

Displaying All Apps on the Home Screen

If you don't want to use a separate Apps screen, you can set your smartphone to show all apps on the Home screen.

To do this:

1. Tap and hold an empty area on the Home screen.

2. Then tap Home screen settings → Home screen layout → Home screen only → Apply.

From there, you can now gain access to all your apps just by swiping to the left on the Home screen.

Indicator icons

You will see indicator icons that show up on the status bar at the top of the screen. Here are some of the icons you will see:

• Since the status bar may not show up at the top of the screen in some apps, you will have to display the status bar, by dragging your finger down from the top of the screen. You can only see some indicator icons if the notification panel are opened.

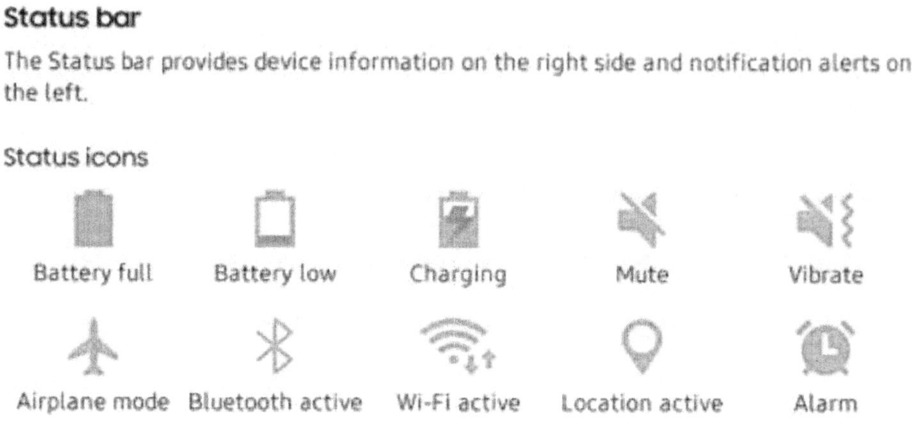

Chapter 10: How to Use Lock Screen Feature

To lock the screen of your smartphone device, press the Power key to turn off the screen and lock it. Also, the screen turns off and locks automatically if you do not use the device for a specified period. You can unlock the screen by swiping in any direction when the screen turns on. But when the screen is off, turn it on by pressing the Power key on the screen.

Changing the Screen Lock Method

You can change the screenolock of your smartphone by launching the Settings app, and then tap Lock screen → Screen lock type, and then choose a method.

When you set a PIN, pattern, passwords, or your biometric data for the screen lock you have chosen, ensure you protect your personal info by keeping others from having accessing to your device. After setting the type of screen lock you prefer, your smartphone will require an unlock code anytime you unlock it.

• **Swipe:** You can swipe in any direction on the screen to unlock it.

• **Pattern:** Drawing a pattern with 4 or more dots, will unlock the screen.

• **PIN:** Enter a PIN with at least 4 numbers to unlock the screen.

• **Password:** Enter a password with at least 4 numbers, characters, or symbols to unlock the screen.

• **Face:** You can also register your face to unlock the screen and set your phone to carry out a factory data reset if you forgot the unlock code and have incorrectly entered it several times in a row and have gotten to the attempt limit.

To do this,

1. Open the Settings app; tap Lock screen → secure lock settings.

2. Then use the preset screen lock method to unlock the screen and then tap the Auto factory reset switch to have it activated.

Adding App Shortcuts on Your Galaxy Phone

The Galaxy phone allows you to add shortcuts for most of the things you like to do on it. You will get to add shortcuts to lock your screen as well as easier access to your favorite apps with no stress.

Adding Shortcuts to Favorite Apps

You are allowed to have a maximum of five apps on the favorite bar. In case you want to make use of this space to add a shortcut, you will need to remove one first.

Remove Home screen shortcuts by:

1. Tap and hold the app to be removed.
2. Select Remove from Home.

Add Home screen shortcuts by:

1. Go to the Apps screen.
2. Select and hold the desired app.
3. Get it dragged to the desired location on the Homescreen.

App with Text Bubble Listing Options Including Remove From Home

In a situation where you have lots of favorite apps, you will need more than the approved five home screen shortcuts. But it's nothing to get worried about as you can still create a folder and get shortcut dragged over each other.

Adding Shortcuts to Home Screen

Get to place all your most-used apps all at once on your Home screen in order by taking the following steps:

On your screen, access the Apps screen by swiping up or down after which you are required to take the following steps:

i. Tap and hold the app until a menu is displayed.

ii. Tap Add to Home.

iii. Tap and hold the app until the Home screen is displayed.

iv. Release app to the preferred location.

v. Tap and hold the app.

vi. Get it dragged to the top or bottom of the screen.

vii. Release app to the desired location.

Adding Shortcuts to the Lock Screen

Two shortcuts can be set up for you to get instant access to your apps from your Lock screen.

To add shortcuts to your favorite apps, you will need to take the following steps:

i. Go to Settings.

ii. Search for App shortcuts and select it.

iii. To set two shortcuts, tap on Left shortcut and Right shortcut.

iv. After setting the shortcut, from the lock screen you can swipe app shortcuts in any direction to open it.

It is to be noted that in case you have set up the lock screen, your credentials will need to be entered before the app opens.

Chapter 11: Notification Panel

When you receive a notification, it will be indicated at the uppermost side on the screen (Notification Bar). To view this, place you finger on the uppermost side of the screen and slide it down. To check the notification details, you can click the on notification message.. Also, you can make use of the below functions in the Panel

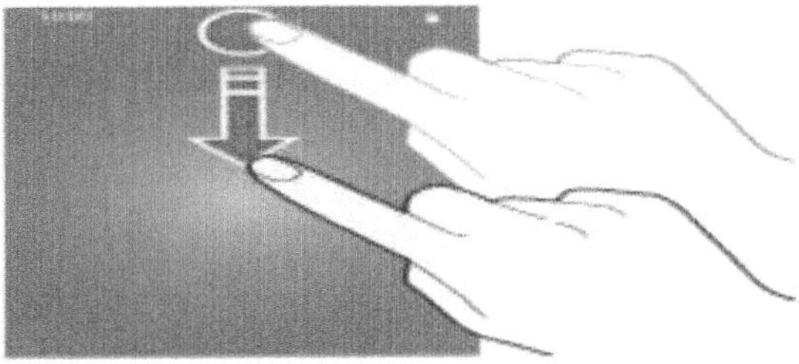

To alter the default feature settings, click on the text below on each of the buttons. To see a detailed info, make a long hold on a button. You can also rearrange the buttons by tapping ⋮ icon button order, Tap and hold a button and then drag it to a different location.

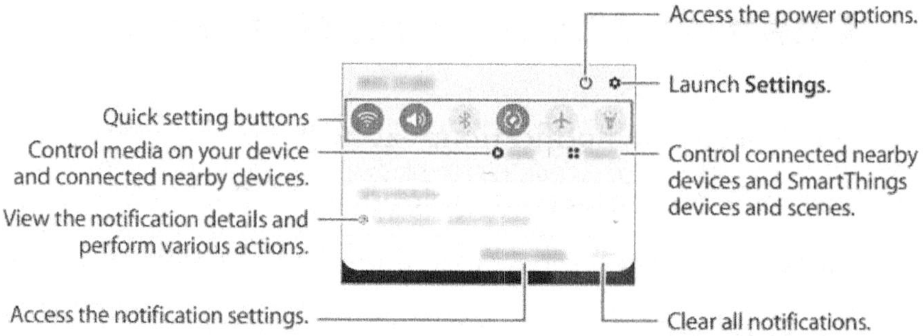

Using Quick Setting Buttons

To activate certain features, Tap quick setting buttons. And view more buttons by Swiping downwards on the notification panel.

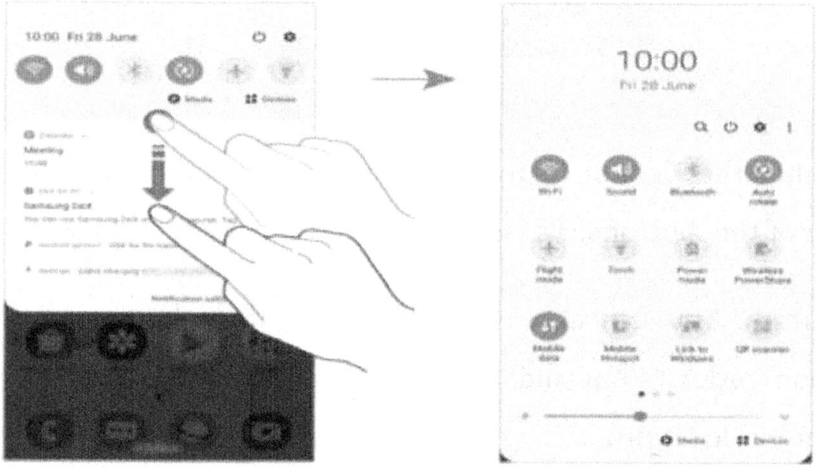

1. Tap the text under each button, to change feature settings.

2. Tap and hold a button to view more detailed settings.

3. Tap ⋮ → Button order and tap and hold a button, and then drag it to another location to rearrange buttons.

Chapter 12: Enter Text in Samsung Galaxy A71

Text can be entered using a keyboard or your voice.

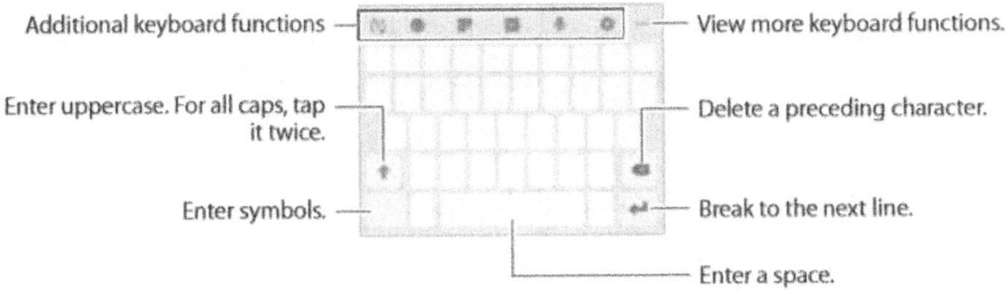

Tap a field to display the Samsung keyboard.

• Tap Show predictive text to view suggested words while you type.

•Tap Show toolbar functions to view all keyboard options.

Toolbar Functions

The toolbar provides additional helpful functions.

○ From the Samsung keyboard, tap Expand toolbar for the following options:

• Clipboard: Access the clipboard.

• Emojis: Insert an emoji.

• GIF keyboard: Add animated GIFs.

• Handwriting: Use your handwriting to enter text.

• Modes: Select a keyboard layout.

• Settings: Access keyboard settings.

• Stickers: Add illustrated stickers.

• Text editing panel: Use an editing panel to help pinpoint text that you want to cut, copy, and paste.

• Voice input: Use Google Voice™ typing.

Configure the Samsung keyboard

Set customized options for the Samsung keyboard.

○ From the Samsung keyboard, tap Settings for the following options:

• **Languages and types:** Set the keyboard type and choose which languages are available on the keyboard. – To switch between languages, slide the Space bar sideways.

• **Smart typing:** Use predictive text and auto-correction features to prevent common typing mistakes. Type by swiping between letters.

• **Keyboard layout and feedback:** Customize the appearance and function of the keyboard.

• **Reset to default settings:** Return keyboard to original settings and clear personalized data.

• About Samsung keyboard: View version and legal information for the Samsung keyboard.

How to Use Google Voice Typing

Instead of typing, enter text by speaking.

1. Tap voice input from the Google keyboard.

2. Then tap the screen and speak directly to your text.

Configure Google Voice Typing

Set customized options for Google voice typing.

1. From the Samsung keyboard, tap Voice input.

2. Tap Settings for options.

• Languages: Choose the language for the keyboard.

• Offline speech recognition: Download languages for offline Google voice typing.

• Hide offensive words: Hide potentially offensive words with asterisks.

Physical Keyboards

Customize options when you have a physical keyboard connected to your device (sold separately).

1. From Settings, tap General management > Language and input.

2. Tap Physical keyboard, and then choose an option:

- Show on-screen keyboard: Show the on-screen keyboard while a physical keyboard is also being used.

- Keyboard shortcuts: Show explanations of the keyboard shortcuts on the screen.

Show Keyboard Button

Show a button on the Navigation bar to quickly switch between keyboards.

1. From Settings, tap General management > Language and input.

2. Tap Show keyboard button to enable this feature.

Autofill Service

Save time entering information using autofill services. From Settings, tap General Management > Language and input. Then tap Autofill service and select your preferred service.

Text-to-speech

Configure your Text-to-Speech (TTS) options. TTS is used for various accessibility features, such as Voice Assistant. From Settings, tap General management >Language and input > Text-to-speech for options:

- Preferred engine: Choose either the Samsung or Google Text-to-speech engine. Tap Settings for options.

- Language: Set the default speech language.

- Speech rate: Set the speed at which the text is spoken.

- Pitch: Set the pitch of the speech.

Chapter 13: Record and Capture Your Galaxy A71 Screen

It is certain that, on your unique Galaxy A71 phone, amazing things will emerge. Therefore you might want to share these amazing things with the latest Screen recorder and screenshot functions which make sharing even easier. What are the things you may want to share?

A good meme, a game you recorded while playing, a video showing different facial reactions, a demonstration of how to create something on the phone, as well as many others.

Note: Not all screens allow recordings or screenshots.

Record Your Screen

However, you should remember that the screen recorder does not function for phone recording, VOIP calls or video calling. In addition to this, owing to piracy constraints, some applications may block the screen recorder function.

Make sure you use the screen recorder feature on your phone for screen recording rather than trying it out with third-party apps. You can always make a clip of anything displayed on your screen with Galaxy A71!

It is equally possible to make a record of yourself during the clip using the front camera.

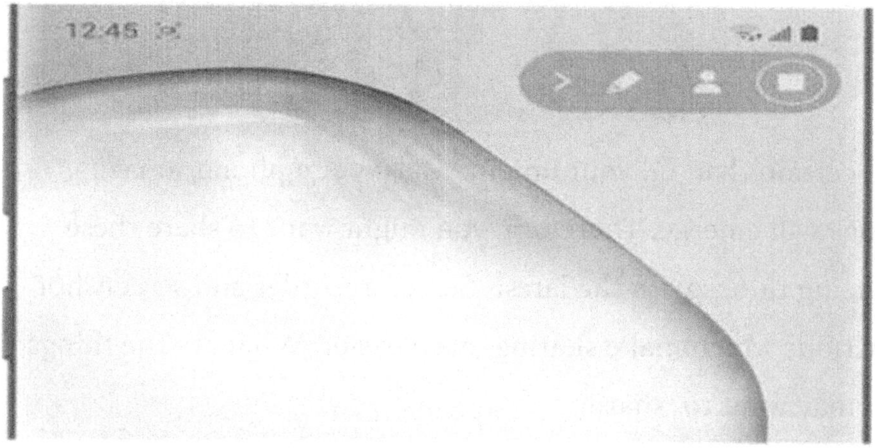

1. With two fingers, flick the screen down from the top to access the Quick Settings panel.
2. Swipe left, after which: press 'Recorder Screen'. The Galaxy A71 will begin to capture anything on the display once the schedule timer stops.

While the screen is being recorded, add to the clip by pressing the front camera icon which appears like a user icon. After pressing it, a small window is created at the top left corner where everything you say or do will be recorded as well.

Click the 'Stop icon' when you have finished capturing. The clip will be stored in Gallery instantly.

To make it more interesting: You can write on the screen while making your clip. To do this:

- Press the 'Pencil icon',
- Pick the color you want and

Access Screen Recorder Settings

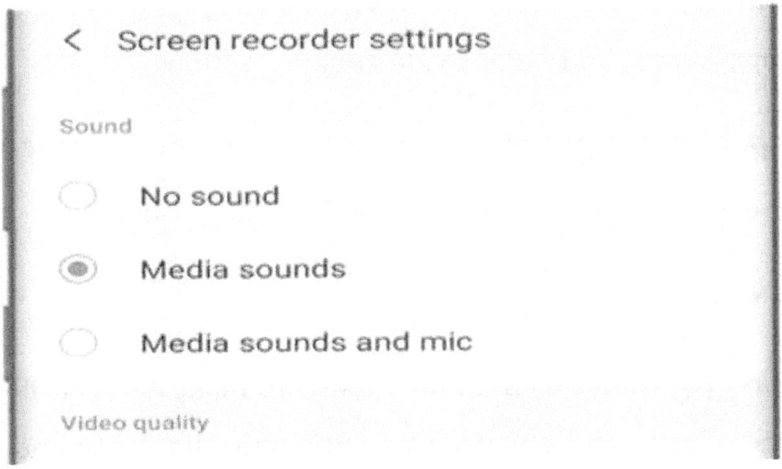

In case you don't want your clip to record unwanted sounds from the background like your own voice, The Galaxy A71 device can make these changes and other kinds to make your video great.

- Swipe down the top screen through dragging with your two fingers to bring out the panel of Quick Settings.

- Flick to the left, then

• Tap and hold the 'screen recorder'.

It will bring out some alternative options for the screen recorder settings:

Sound: Choose the type of sound for the clip to record. The options are: 'Choose from No Sound', 'Media Sounds', or 'Media Sound Mic'.

Video quality: select the required video quality. You will find the 1080px, 720px, and 480px video quality options.

Size of the video selfie: This will help you adjust the size of the window created at the top corner for recording yourself.

Take A Screenshot

The Galaxy A71 device uses different keys to make a screenshot. As you now know, the new Side key integrates the Power and Bixby key as one. Nevertheless, as you try to adapt to this change, the good thing is that with one hand, you can now capture anything on the screen.

Tap the Side key together with the Volume down button simultaneously to capture a screenshot. You will get a screen flash after which a list will be seen at the bottom of the screen. The list gives you some alternatives that include:

Screenshot made with options shown in a Scroll capture bubble

Scroll Captures**:** Take a lengthy screen capture to record the whole thing. This configuration is accessible only when it is possible to scroll.

Draw: With this, you get to edit the shot before sharing it with friends or saving it on your device.

Add Tag**:** For quick access, put a tag on the shot and save it.

Share: Share with friends or use on other applications.

Set Your Phone to Do Not Disturb

In case you want to read throughout the night for a test, you probably don't want to get distracted. And yes, you can make this adjustment on your device. The Do Not Disturb option helps you to concentrate on what you want to achieve. If you really need some private and quiet time, the option disables all the notifications on your phone.
In addition to this, the option helps you to switch on or off your phone at specified times, such as a business lunch or a thanksgiving dinner.

However, if you don't want to miss an important call or text, you can sometimes set some notifications as special cases.

Keep in mind that the type of internet service provider and software version of the available screens and configurations may differ.

Turn Do Not Disturb on or off

Initiate the Do Not Disturb option when you need some rest. So you can enjoy the tranquility. To do that:

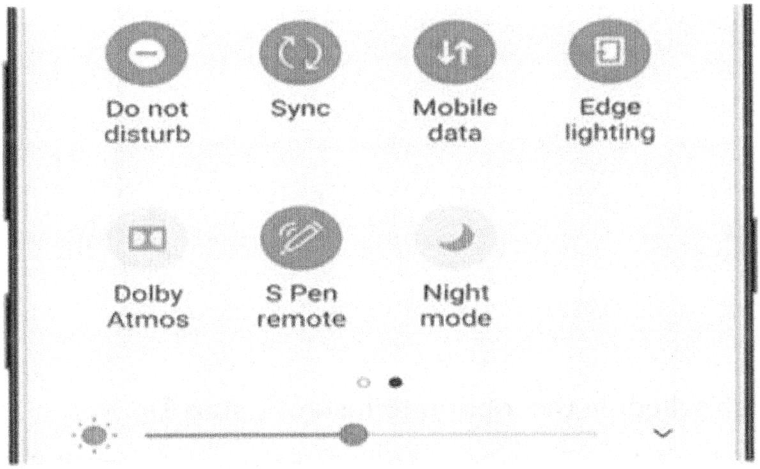

1. Drag down the Quick Settings with your two fingers.
2. Flick to the left and Tap the 'Do not Disturb' sign to switch it on or off.

Another method is to go to Settings, look for the **Do Not Disturb** icon and tap to pick it. To switch it on or off, tap the button next to 'Do not disturb'.

Do Not Disturb Schedule

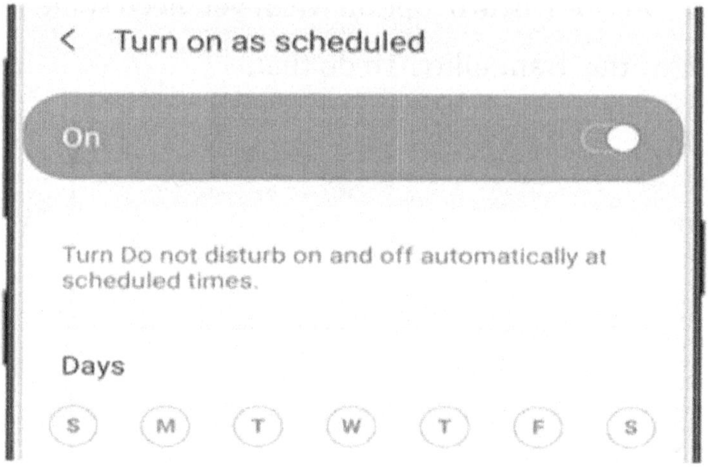

You could also schedule the option to instantly start Do Not Disturb, if you're going for an important conference. You will find the '**Do not Disturb**' from the Settings. Press '**Do not disturb**', then tap '**Turn on as scheduled'.** Tap the screen switch, then pick and modify the option to suit your need.

Set Up Exceptions

You can personalize the settings to enable exclusive notifications especially if you are expecting an important call while the 'Do not Disturb' option is on.

Find and pick 'Do not disturb' from Settings. Press the 'Do not disturb' one more time, tap 'Allow Exceptions', then select from the choices listed.

Chapter 14: Customize Accessibility Settings on Your Phone

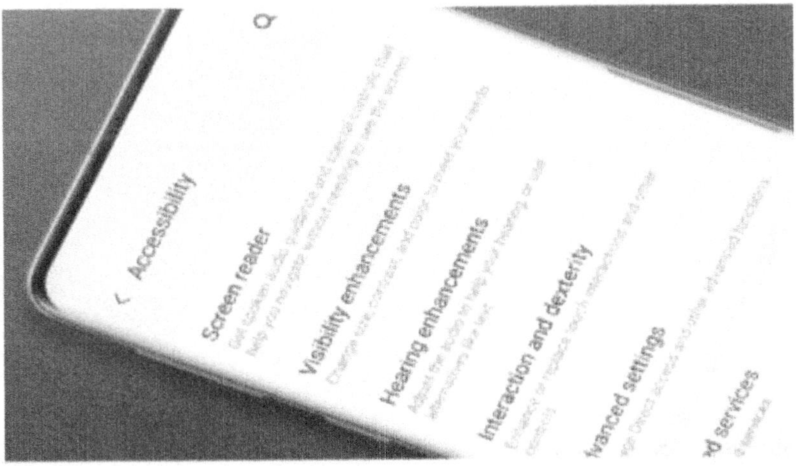

Your device is willing to make these adjustments for you. There are settings that can help people with varying disabilities can use. For instance, those with constrained vision, hearing deficiency, or interaction and dexterity challenges. For hearing deficiency, if the Galaxy A71 recognizes a sudden sound, you can configure it to receive texts aloud or make a flash. For people that are not great at dexterity, there are also other settings and customizations available.

Also keep in mind that internet service provider and OS version of the available screens and settings may differ.

Available Accessibility Settings

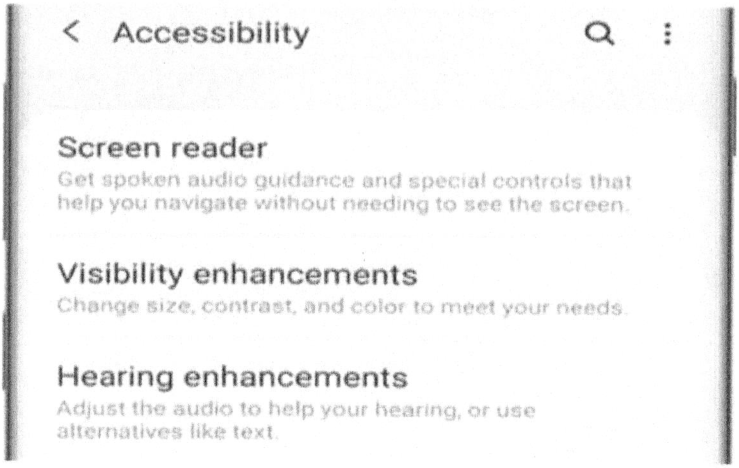

Go to Settings, Scroll over to 'Accessibility' and tap. It will bring out the following function:

Screen reader: This option is a Voice Assistant which gives you voiced instructions anytime you press anything on your device as it also gives an informative guide on how to use the phone. Other settings such as text-to-speech, noise feedback etc can also be customized.

Visibility enhancements: highlighting fonts, turning on negative colors, removing animations, and many others.

Hearing enhancements: Here, sound sensors to get notifications when, around your environment, your device hears sounds like the doorbell or a child weeping. You can silence all sounds on your phone with this option.

Interaction and dexterity: Improve or change touch interactions and other controls.

Two other features: Universal Switch and Assistant menu helps to improve this accessibility option. The former uses custom buttons to control your device while the latter switches on functions that assist users with decreased dexterity.

Advanced settings: You will find different advanced functions here like managing the Direct access feature to open accessibility settings.

Installed services: Here, you will find all your downloaded Galaxy apps from Play Store for accessibility services

Chapter 15: Add or Remove Edge Panels on Your Galaxy Phone

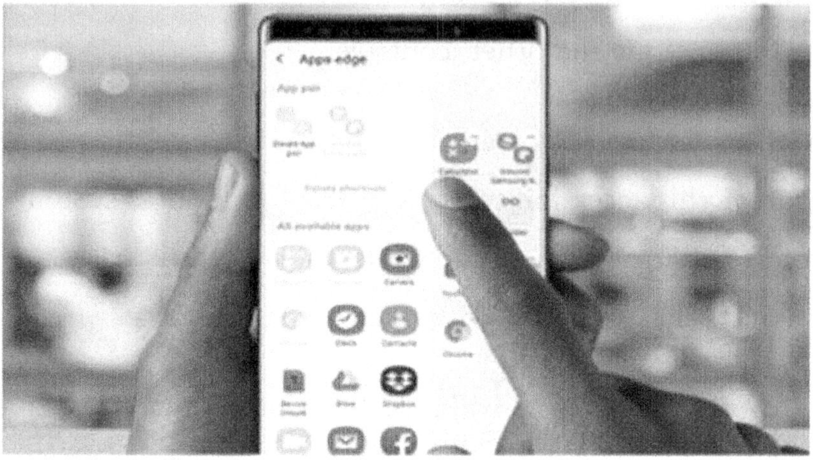

You may be finding it really difficult to locate some apps since there are so many of them on your device. Do not worry, Edge Panels helps you overcome that fast with ease. It functions like a speed dial. Being flexible, Edge Panels can be added, removed or installed whenever you want it.

Add or Remove Edge Panels

Simply add the app you frequently use to your Edge panel for fast access. For instance, apps to check our favorite sports team or stock report apps you want to follow. And you have as many Edge Panels as you want but not more than nine.

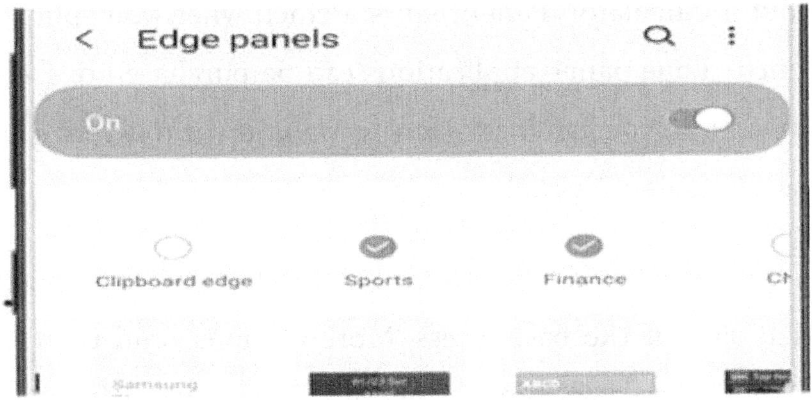

To open the Edge panel, Flick to the left and then tap the Settings icon. Choose or delete the edge you want based on their importance.

Download New Edge Panels

Would you like Google Maps at your fingertips?

How about a Calculator Panel that is a cinch when you split the tab at lunch? Edge panel applications can be purchased or freely installed. With it, you can have Google Maps at the touch of a key as well as a Calculator panel.

To open the Edge panel, Flick to the left and tap the Settings icon on the left side at the base. Press 'More Options', and then tap 'Galaxy Apps'. Agree to the Terms and Conditions after going through it, and then click 'START'.

To install it, tap the Edge panel app you want and follow the instructions displayed for you.

Edit Edge Panels

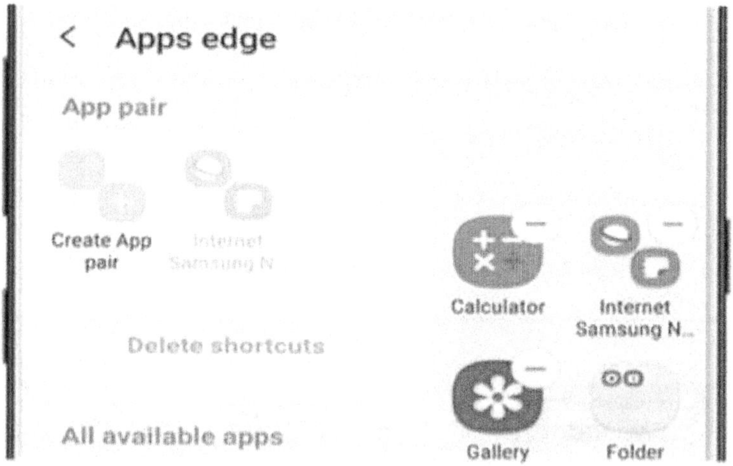

Configuring the Edge Panel is not all about picking and deleting apps alone. Each panel can also be edited. For instance, the most important app can be shifted to the top of the panel.

To open the Edge panel, flick to the left and tap the Settings icon. Press the 'Edit' function which you will find below the edge you want to change. You can even generate pairs of apps to facilitate multitasking.

Edge Panel Handle Settings

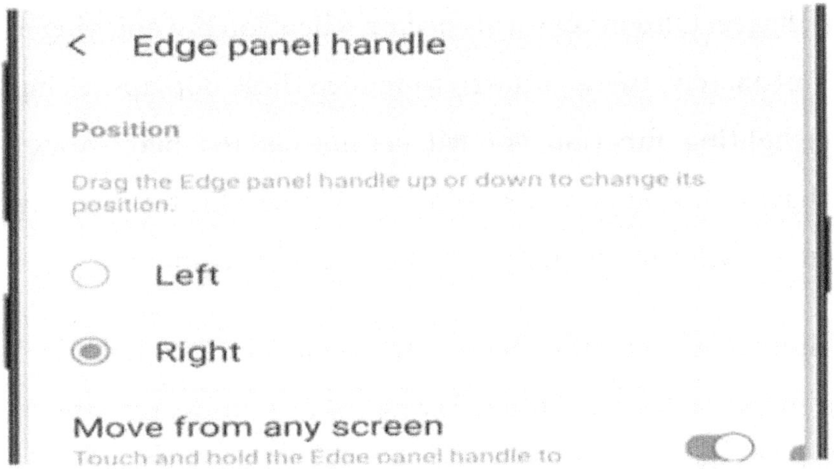

You might not easily see the handle of the Edge panel displayed on the side of your Galaxy A71 device. This is because of the handle's transparency. However, this can be reconfigured together with the location and size of the handle.

- Open the Edge panel by swiping left and tapping the Settings icon at the bottom left of the screen.
- Go to 'More Options', then
- Press 'Edge panel handle.'
- Follow the instructions and make the changes you want.

Use Edge lighting on your Galaxy phone

It is very annoying when messages pop ups and blocks the screen of your device. Often times, this occurs when you are enjoying a game. Fortunately, there is a way to reduce these pop ups. Using the Edge lighting function, you will get notified in a better way rather than annoying pop ups. When you've set this feature, any alert you get will simply light up the corner of the phone.

Note: Internet service provider and OS version of the available screens and settings may differ. Try not to use this feature for third-party apps.

Edge Lighting Settings

It is always better to use Edge lighting in areas that require your full attention. To set it up;
Go to Settings > search for '**Edge Lighting**' and press it > Press the desired switch options.

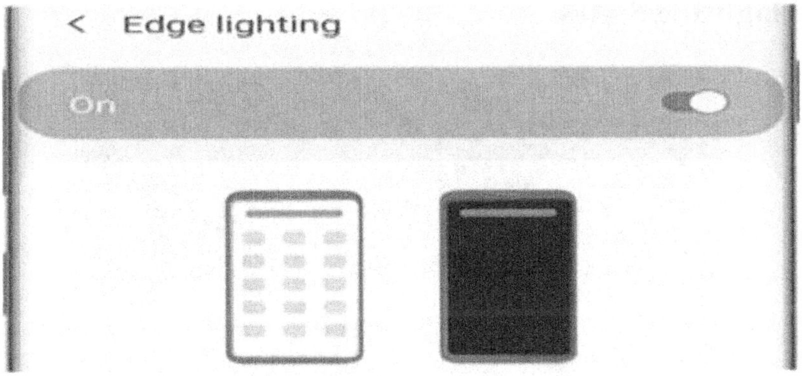

You can personalize the Edge lighting setting if you want to use it for specific times. To do this

Open the Edge Lighting option from Settings > Press 'Show Edge Lighting' and it will provide these options:

1. When screen is on: this will light up the edges only when the screen is on.

2. When screen is off: this will light up the edges only when the screen is off.

3. Always: this will always light up the edges whether the screen is off or not.

4. Keep in mind that the default settings is set at 'Always'

Edge Lighting Styles

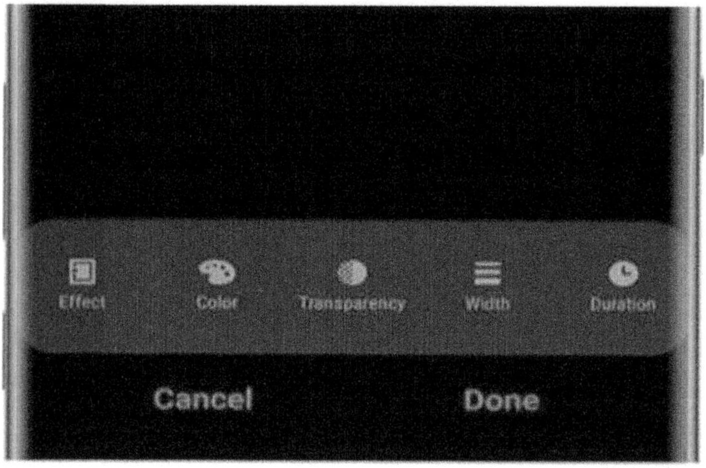

There are other configurations to do on the Edge Lighting feature. Apart from barring notifications at specific times, you can also change the color and style of the lights.

- First and foremost, open the Edge lighting from Settings.

- After this, scroll down to and press 'Edge lighting style' and then

- Choose your preferred styles.

- Press 'Done' when you are through with your selections.

Here are the different styles you are provided with:

1. Effect: You are provided with varying designs like Basic, Multicolor and others

2. Color: Pick the hue light you love and you can pick a particular color for every app by touching 'Custom color'. However, this cannot be done with Glitter Design.

3. Transparency: Pick from high transparent intensity or low.

4. Width: You can pick from narrow to wide.

5. Duration: This is how long you want the light to appear: pick either long or short.

Edge Lighting Interaction

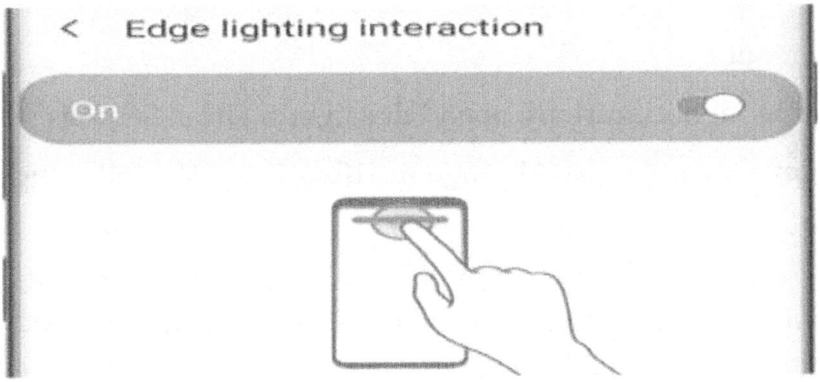

In addition to these configurations, it is possible to interact with any light-up alert you get just as you can do for the usual notifications.

To do this: Open the Edge lighting from Settings > Press 'Edge lighting interaction' > Press the switch to initiate the configuration.

Edge Lighting Notifications

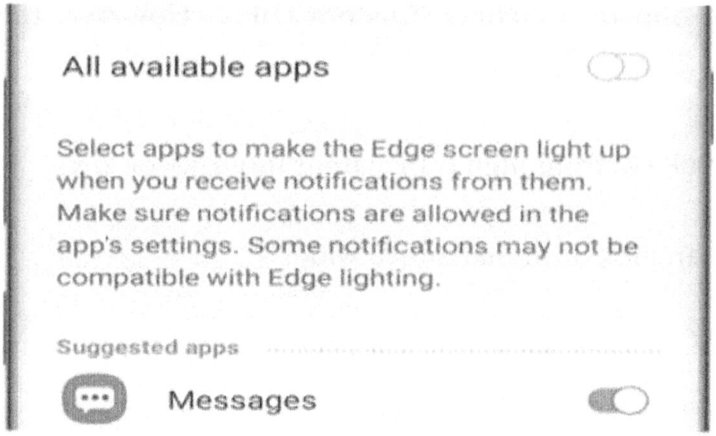

If you are looking for a special type of alerts for a specific app which you can use from time to time, Edge lighting features can help you out.

To get special notifications, open Edge lighting from Settings > scroll down to and press '**Manage notifications**', the following options will be displayed.

1. **All available apps**: click on this switch if you want one notification for all apps
2. **Desired apps**: click on this switch if you want one notification for a particular app.

Keep in mind that Edge lighting does not function for all third-party apps.

Chapter 16: Adjust Your Galaxy Phone's Display Settings

Don't worry about the capacity of your Galaxy A71 device because it can perform so many daily activities by using different screen options like the full screen. Regardless of the type of activity you perform on your phone, the screen features will help your phone adjust to it.

Customize Your Font Size and Style

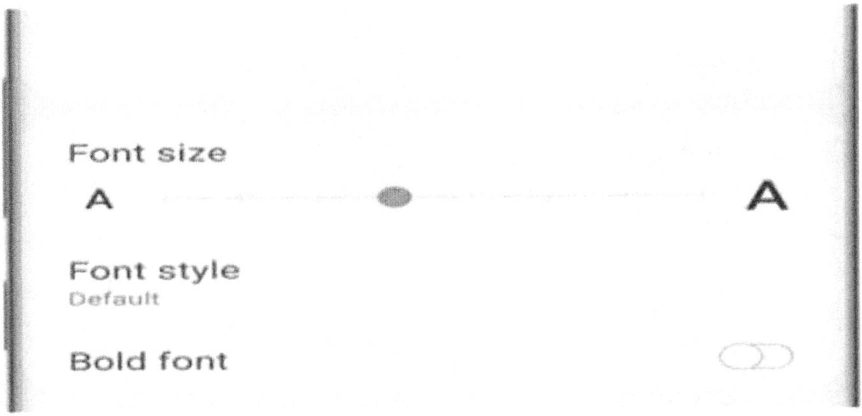

Let's say you are surfing the net for some valuable information but it seems your eyes are being affected. What is the problem and how can you use the screen features to overcome this? First and foremost, the problem could be the size of the fonts and all you need to do to overcome this is to change the size of the font. To do this:

- Go to Settings
- Press 'Display'. It will bring out a list of options.
- For font size adjustment,
- Press 'Font size and style', and then
- Change the size to suit your needs.
- If you want to adjust the entire screen, making it smaller or bigger:
- Press 'Screen zoom', a slider will appear and so,
- Flick the slider to suit your needs.

Turn On Full Screen Apps

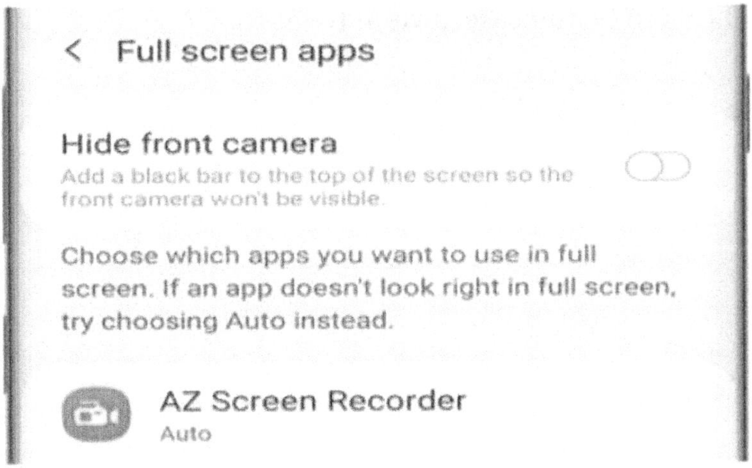

In addition to screen size or font size, you can adjust the screen space of your game apps or make videos look bigger on your device. However, you need to confirm if the app is compatible with full screen.

To do this with Full screen apps:

- Go to Settings.
- Look for and press 'Full screen apps'. It will bring out apps with Full screen functionality
- Scroll down and look for the app you want to configure and then,
- Click the button beside the app to enable full screen.

Note: Not all apps are compatible with the full screen mode but if you find this to be the case, switch to auto.

Choose a Screen mode

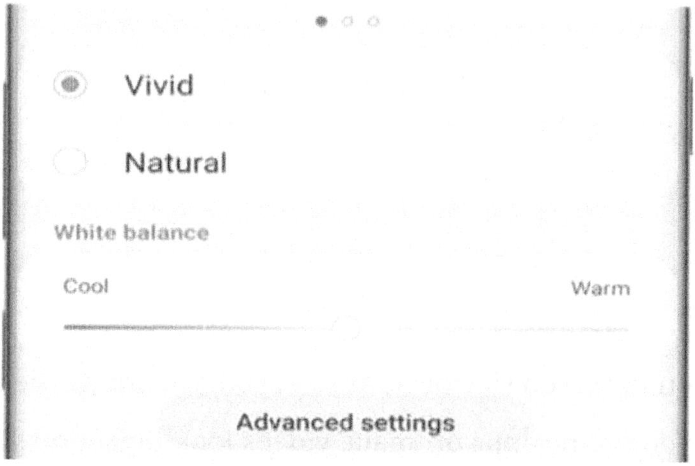

Change the display of that funny looking image by utilizing the Screen mode settings. To do this:

- Go to Settings
- Click 'Display', and then
- Press 'Screen mode'. Next,
- Pick the screen mode you want.

Nevertheless, until the Blue light filter is switched off, some screen modes may not function.

Note: The availability of screen modes depends on the type of phone version.

Adjust Your Screen Brightness

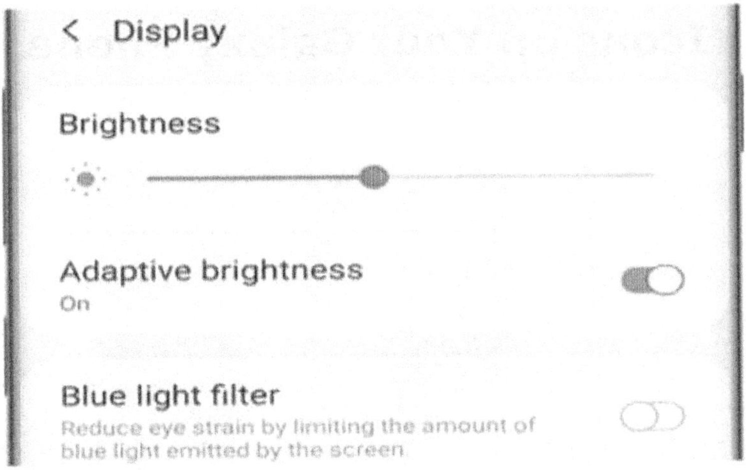

You must have discovered that it is difficult to see your screen on sunny days. This should not bother you because you can change the brightness or just allow your Galaxy A71 device does it automatically. To do this:

- Go to Settings
- Press 'Display', and a slider will appear and then
- Swipe the pointer to a position that suits you.

However, you will have to make use of the Adaptive brightness functionality, to make this feature automatic. When you open '**Display**' from the Settings, simply press the Adaptive brightness feature and switch it on.

Chapter 17: Customize Themes and Icons on Your Galaxy Phone

Depending on what you love most, you can change the theme of your device by customizing the following features: icons, wallpapers, and themes to fit your personality. For easy access, download the themes on another phone if you have a Samsung account. However, keep in mind that the type of internet provider, Android and phone version will determine the type of adjustment you get. But, note that you will not be able to remove the icon frames when you've updated the Pie OS update.

Download themes

A fresh theme makes your phone feel brand new as it spices it up. Fortunately, you can download one. To do this:

- From the home screen, long press an empty area.
- Press 'Themes' from the options provided
- Scroll up to check the theme you would like
- Press 'Download' on the preferred theme
- Press 'Apply'
- Press 'Apply one more time for confirmation.

Change your theme

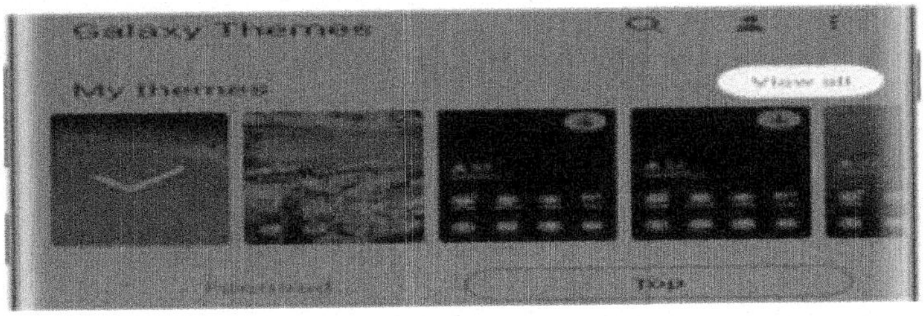

You can also change your theme if you are tired of seeing or just want something fresh. To do this:

- From the Home screen, long press an empty area to open the themes
- Press 'View all' to check out all your previously downloaded themes.
- Pick a theme and check out the details and then
- Press 'Apply' twice to complete the process.

Download icons

Your phone icons reveal a lot about you and one of them is this: your favorite hobby. Tell people about you by installing amazing icons you love. To get this done:

• Open the themes option by long pressing an empty Home screen.

• Press 'Icons'

• Scroll up to check out the icons

• Press 'Download' on the chosen icon

• Press 'Apply'

• Press 'Apply' once more for confirmation.

Change Your Icons

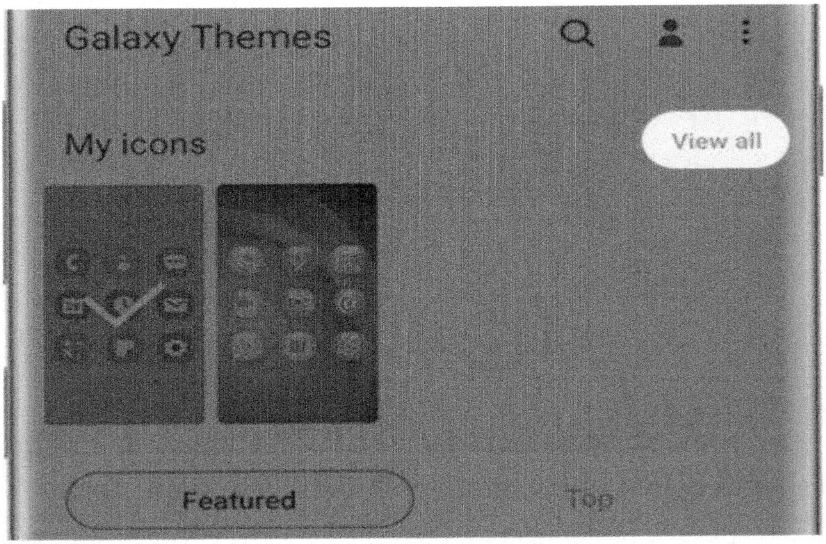

But then, you could get tired of seeing the same icon over and over again. You can simply change the icon to another pre-installed one. To dothat:

- Open the themes by long pressing the empty Home screen
- Press 'Icons'
- Press 'View all'
- Check out the one you like for the pre-installed icons
- Press 'Apply' twice for confirmation

Delete Themes

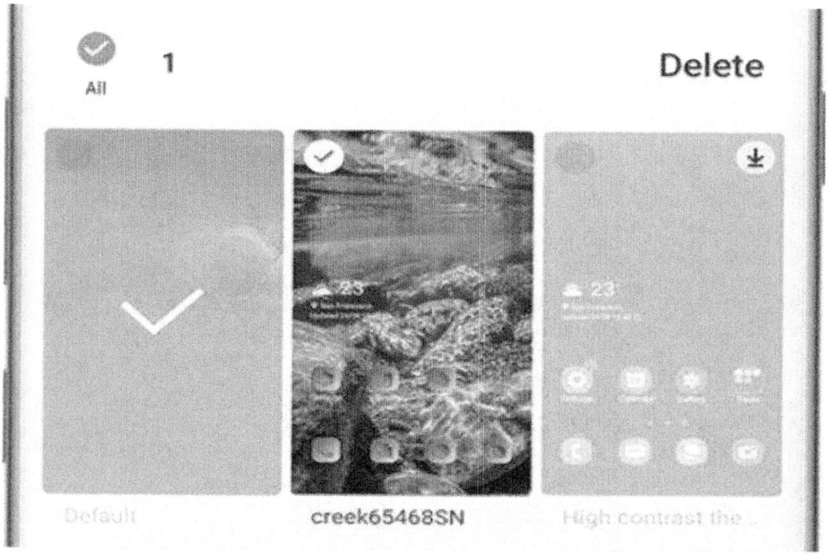

You should not worry about downloading the wrong theme. It happens. Fortunately, in just a few clicks, you will get rid of it.

- Open the themes by long pressing the empty Home screen
- Press 'View all' to fish out the unwanted theme.

• Press 'Delete' twice to confirm the operation.

Delete icons

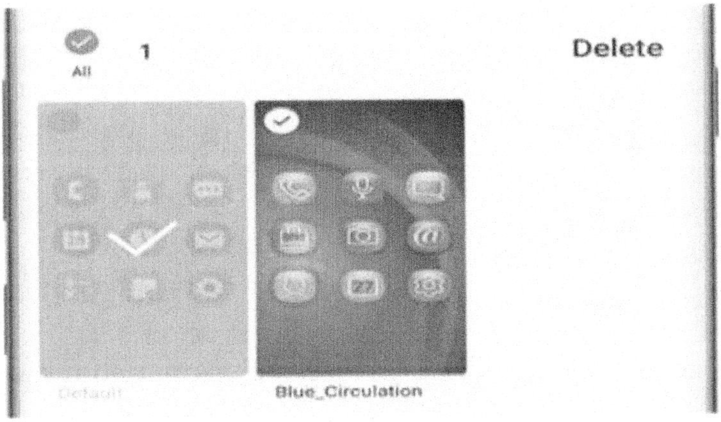

Provided you see an icon over and over again to the point that it irritates you; you might want to just delete it. It's easy:

• Open the themes by long pressing the empty Home screen

• Press 'Icons'

• Press 'View all' to find the boring icon

• Press 'Delete' twice to get rid of the icon.

Recover Purchased Themes

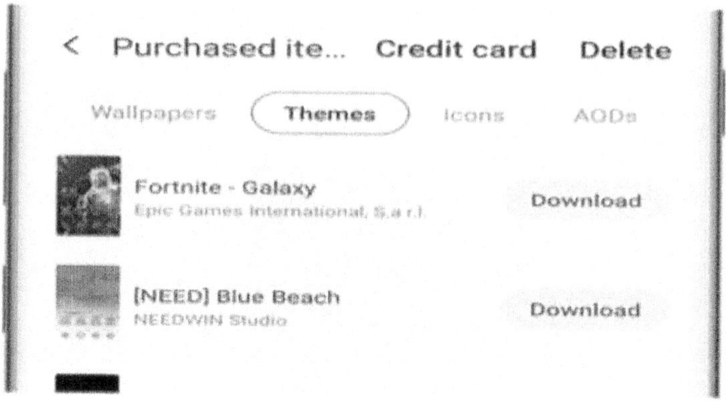

If you used a Samsung device before, set up an account where you paid for some installations, then, you shouldn't bother about setting up another account. Interestingly, you shouldn't pay again for those items in your former account. Here is how to integrate the items in your Samsung account for free.

- Open the themes by long pressing the empty Home screen

- Press 'My Page' which looks like a human icon

- Press 'More Options'. It will bring out different options and then,

- Press 'Purchased items' and all the previously purchased themes will appear

- Press 'Download' on the theme you want and that's all.

Note: this can be done as many times as possible and you can also use the same process to get all the previously purchased wallpapers and icons.

Customize Your Phone's Font Settings

Spice up your Galaxy A71 device by changing the font size and style. Maybe it is too difficult for you to read or you want something new, whatever your reason is, you can spice things up with fresh downloaded fonts. However, keep in mind that the type of Android version, phone version or internet connections will determine the type of screen display or settings.

Change Font Settings

Select a perfect font that suits your lifestyle. You can start with the size of the font. To do this:

• Go to Settings and search for Font size and style

• Press 'Font size and style' twice and then you will see a pointer which you will swipe right or left to adjust the font size.

• Press the switch beside the bold font to use this feature.

• Press 'Font style' to get another font or download fresh ones.

However, keep in mind that not all apps allow big fonts and those app will automatically pick normal font size.

Download New Font

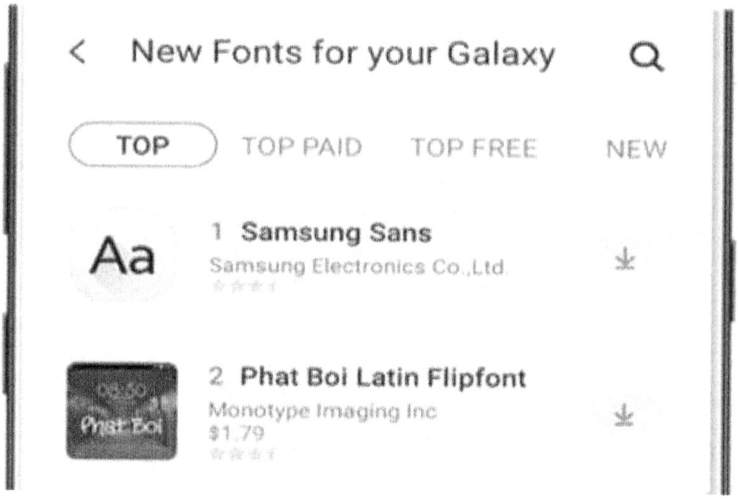

Change old fonts when they become boring or when they seem out of place with your current mood. To do this:

• Open Font size and style from the Display option under Settings.

• Press 'Fonts size and style' once more

• Press 'Font style' from the list of options

• Press '+ Download Fonts'. This will open the Samsung Galaxy store where you can pick the font you like.

• Press 'Install' next to the font you prefer.

High Contrast Fonts

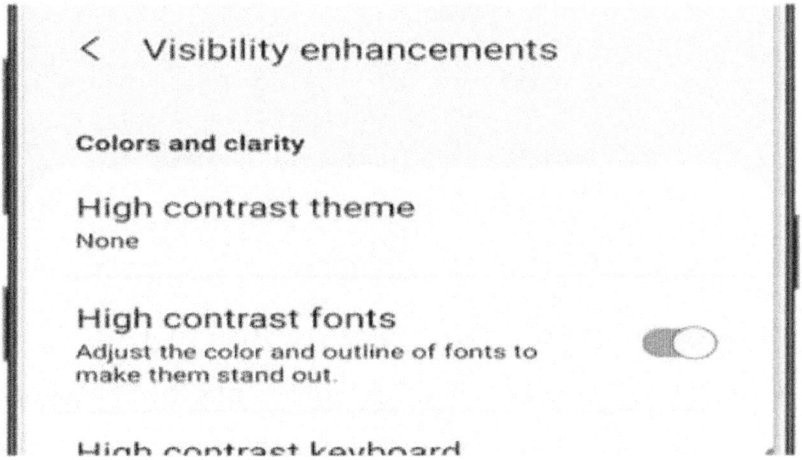

What is the Contrast feature used for? It is simple: it is used for adjusting the look of your fonts. If you want the fonts to stand out better, increasing the contrast is the way to do it.

• Go to Settings and look for 'Visibility enhancements'

• Tap the pointer beside it to switch it on or off the High contrast.

Chapter 18: Use the Samsung Messages App on Your Galaxy Phone

The Messages app on the Galaxy A71 is a great tool for keeping tabs with all your contacts. The app makes photo sharing and sending emoji very simple. In addition to this, you can quickly say hi, easily mute chats and even schedule important messages for specific contacts.

Set Custom Tones for Message Threads

There are different contacts on your phone and they all represent different things to you. Make this difference known by choosing different ringtones for different contacts. This way, you will know when someone is texting you without looking at the screen.

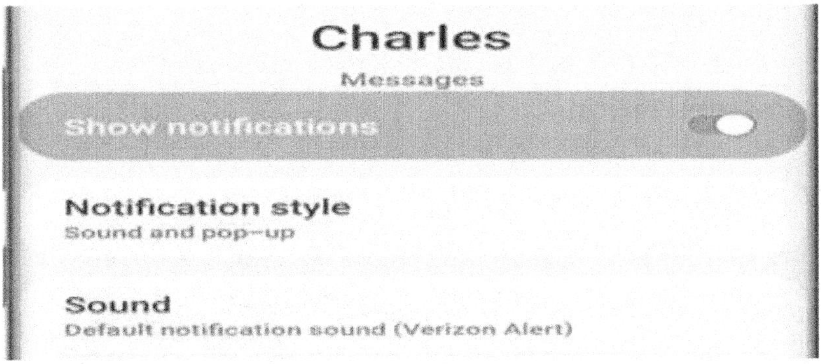

To do this:

- Open Messages from the menu
- Touch the conversation you want to customize.
- Press 'More Options'
- Press 'Custom notifications'
- Touch the switch at the top of the screen displayed.
- Press 'Sounds' after which you pick the alert sound for that conversation.

Note: Your phone carrier and network will determine the notification settings you get.

Message Customization

Samsung is known for stylish designs that make your device unique. Therefore, you can change the outlook of your message app on your Galaxy A71. You should change the theme of your phone to start with and then, change the font size and style as well.

Change the Font Size

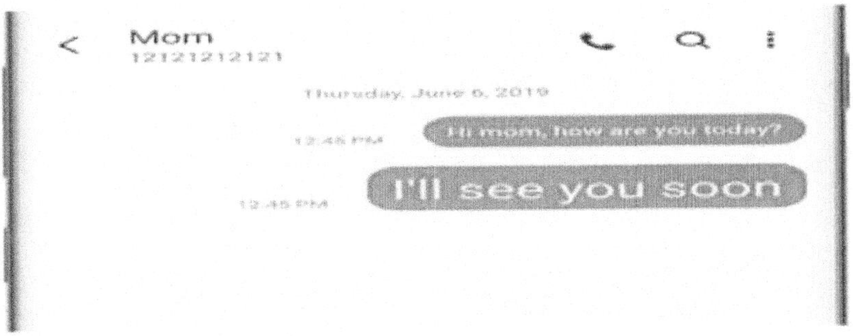

To ensure that you don't miss any word on the message sent to you which is caused by small font size, change the font size of each message. This will not affect the phone's font size. To make this happen;

- Go to the message app
- Select the message you want to see

Using two fingers, zoom the message by dragging the screen in opposite directions.

This makes the font bigger. Equally, every adjustment you make will only affect messages from the app alone. However, the selected size will be stored until you change it back.

Create and Schedule A Message

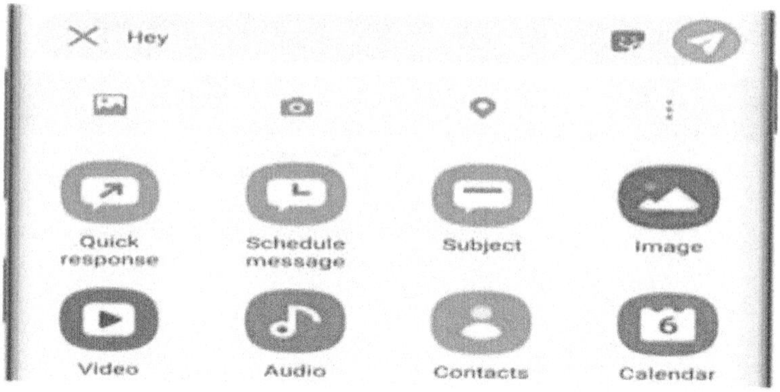

A list of icons for text message options

What can the Messages app do?

Basically, the app helps you to create messages I.e greeting texts. The app can also schedule messages which will be sent later. Here is how to create a message:

- Go to 'Messages' from your phone menu
- Press 'Compose'. You will find at the bottom right of the screen.
- Pick the contact you are sending it to
- Press 'Enter'
- Create your message and then
- Press 'SEND'

If you wish to send th message later, you will have to schedule it. Here is how to schedule a message:

After composing the message,

- Press 'Add' which is in form of a plus sign

- Press 'More' and then from the options

- Press 'Schedule message'

- Pick the date and time you want your message sent. Note that you can pick a year as your scheduled time.

Other options you get when you press '**More Options**' include the following; Quick responses and Subject features. You can also attach images, videos, audio, contacts, Calendar events, and Samsung Notes to the message.

However, keep in mind that incomplete messages are automatically stored under 'Drafts'

Messages App Settings To Know

To utilize your Messaging app very well, there are some important tools you should not ignore. These are:

The Search icon: The search icon can be used to find old messages you might want to see again.

Delete: This will help you get rid of unwanted messages. To use this feature, long press the unwanted message and hence, press 'Delete' which you will find at the bottom right. Press it once more for confirmation.

Star messages: This is will help you keep track of important messages. To star a message, long press the favorite message and press '**star message**'. If you would like to see the important message: go to 'Conversations' on the Messaging app menu > press the three vertical dots which represent 'More Options' > press 'Starred messages'

Block messages: With this feature, you will not only be able to delete unwanted message but block the number used to send it. If you want to block a user that sent an annoying text to you, pick '**Block number**' when a pop-up shows up on the screen. Press 'Delete' afterwards.

To see the numbers or messages you've successfully blocked simply: go to '**Conversations**', press '**More Options**' and then press '**Settings**'. From here, touch '**Block numbers and messages**'

Update the app: This will help you keep track of new updates to use on the device. To find the new updates: go to '**Conversations**' > press '**More Options**' > press '**Settings**' > press

'**About Messages**' > press '**Updates**' provided there is an update available.

Mute Or Unmute A Conversation

Message notifications are sent to your phone based on its default settings.

This can be annoying especially when you are busy but you can change the settings. To do this:

- Go to your Messaging Apps
- Pick any conversation
- Press the three dots that represent 'More Options'.

This will bring out a list of options but you have to pick between '**Mute Conversation' and 'Bell Icon**' depending on your phone model. Follow the steps to unmute as you will use the '**Unmute Conversation**' or '**Bell Icon**' to return to default settings.

Change the Profile Picture

Edit Contact profile screen

Customize your Messaging app by adding a profile picture. This will give your message a professional look. To set this:

Go to '**Contacts**' from the menu feature. You will see your profile above the list.

- Press your profile icon.
- Press 'Edit'
- Press the icon that will redirect you to your pictures and then
- Pick a picture file from the album
- From the selected album, pick the picture that suits you.

Other image Settings

1. You can resize the image by press-dragging it and the press 'Done' to confirm.

2. Add tags to the image and press 'Save' to complete the process. The interesting part is that Samsung users on your contact will be able to see your profile picture.

However, you have to activate profile sharing on your phone to make it visible to other Samsung users. Custom icons can be used as profile pictures

Chapter 19: Control App Notifications on Your Phone

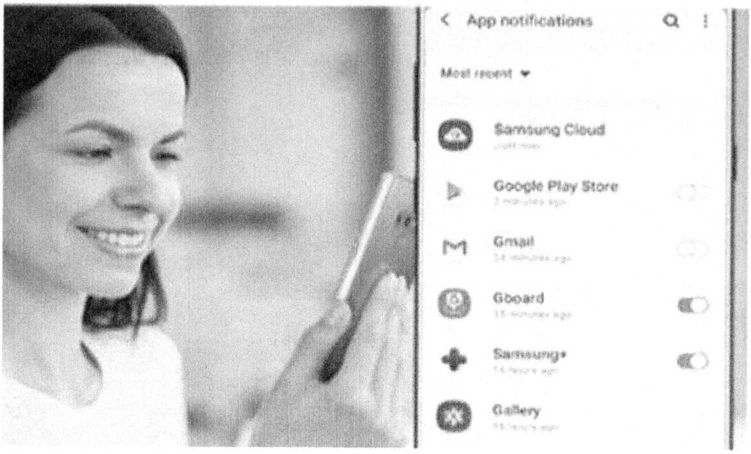

Annoyed that old and boring apps are still sending you alerts? Change all that as you select the favorite apps that you want to get notifications from while you stop alerts from unwanted apps. In addition to this, you can change the type of alerts you want like vibrations and sound.

Note: the type of screen and settings you get is based on your internet service provider and phone OS.

Block App Notifications

Make sure unwanted apps do not send you alerts again and you can do this from your settings.

- Press 'Notifications'
- Tap 'See all'
- Tap the pointer on the unwanted app to switch it off or on.

Customize App Notifications

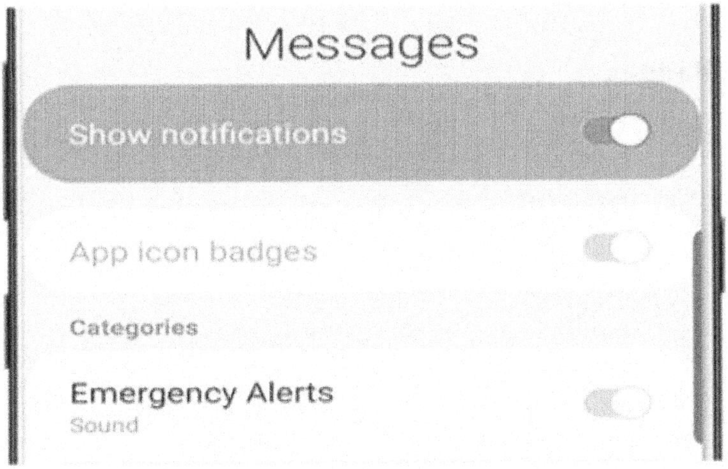

If you want to quickly identify which app sent you a notification, you can pick a unique sound for a particular app.

However, remember that different apps have different options.

To set this up:

- Go to 'Settings' from your phone menu
- Press 'Notifications'

- Press 'See all'

- Tap the app you want to configure

- Switch the notification on or off by tapping the pointer.

Follow these steps if you want to configure the notification of other apps.

App Icon Badges

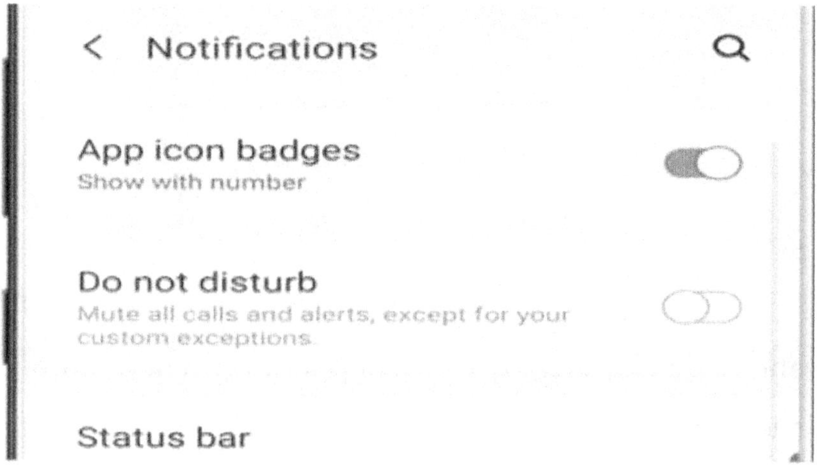

When you get a notification, the little dots you see beside your app are known as App icon badges. If you don't want these dots, you can easily remove it.

- Go to 'Settings'

- Press 'Notifications'. Under the options you get,

• Tap 'App icon badges'

• Press the pointer you find in the feature to turn it on or off.

If you prefer just two dots or one rather than the three, you can change it as well

Status Bar Settings

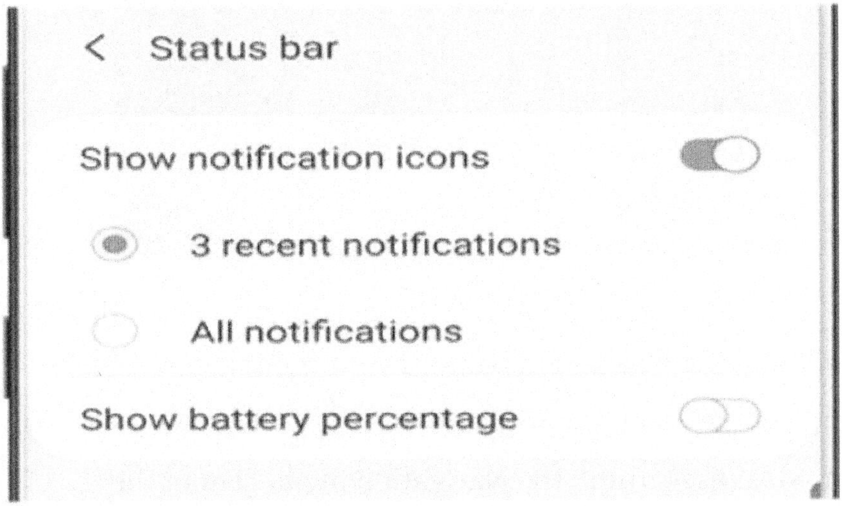

Sometimes your status bar is filled with so many notifications making it annoying. It is possible to hide these alerts or show some.

- Press '**Notifications**' after you've opened the settings menu
- Press 'Status bar'. You can now change the setting here to suit your need.

Chapter 20: Customize the Navigation Bar on Your Galaxy Phone

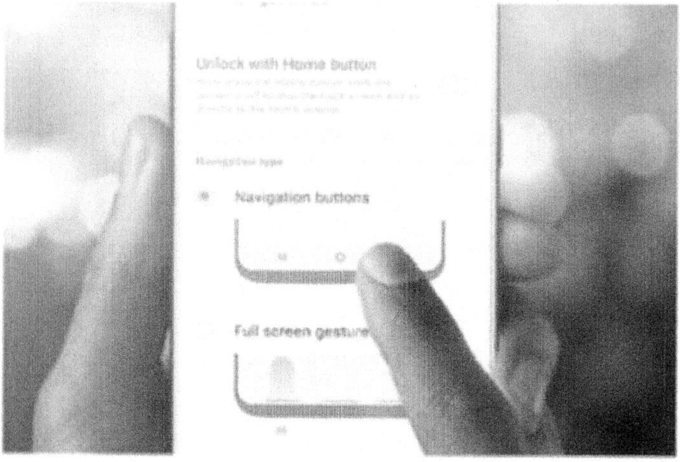

You will find the Navigation bar at the bottom of your phone screen. It is the menu that directs you to anywhere in your device. Most users think the Navigation menu cannot be reconfigured. That's not true. The layout and order can be personalized even as you can totally remove the menu making use of gestures.

Note: The internet service provider and phone version or OS will determine the type of screen or Settings you get.

Change the Navigation Bar type

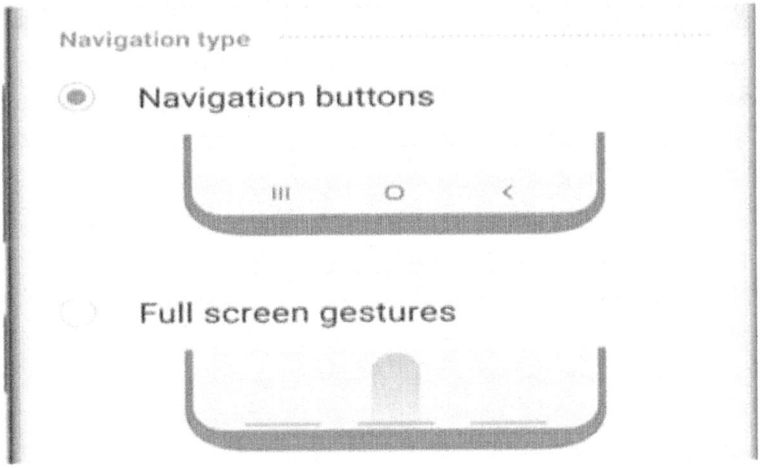

Navigation bar type

The Navigation bar is the menu that directs you to different parts of your phone. In old phone models, the menu is the fixed buttons at the bottom of your phone. To reconfigure your navigation bar, you can start with the Full screen gestures to get a full screen design. This will remove the navigation bar while you swipe up to bring it back when you need it. To hide them back, you use the Gesture hints.

- Go to 'Settings'

- Locate and pick Navigation bar

- Press 'Navigation Bar' once more and then pick either Navigation buttons or Full screen display.

Note: Press the switch beside the Gesture hints to turn it off. This is to hide the full screen gestures lines totally.

Navigation bar Button Order

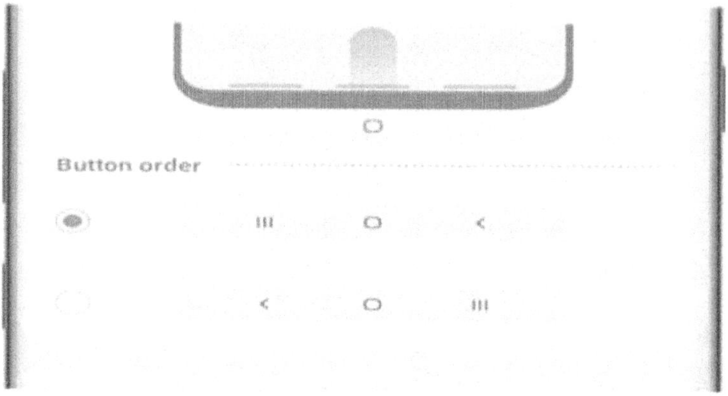

The buttons on the Navigation bar has a default order that can be adjusted. To do this:

- Pick the Navigation bar option from the settings menu.
- Press it once more.
- Find '**Button order**' and finish the configuration.

Navigation Bar display issues

It is puzzling that the Navigation bar will not be seen when games are on and it refuses to go when you try to hide it.

What is the reason for this and how can you resolve it? The fact is this: The Navigation bar can only be hidden on Android Oreo version and not Pie OS. Ingame Navigation bar displays Game tools, Recents, Home, Back, and Lock options. Note that the Navigation bar, although switched on, could disappear when you use third-party apps. This is common in video and games. If you want to see the bar while you are playing games, just swipe up to reveal it.

Chapter 21: Facial recognition security for your Galaxy phone

There are different types of security system that comes with your Galaxy phone, you just have to go by the one of your choice. One, is the Facial recognition, a secured security method that lets you secure your phone by scanning and registering your face.

Facial Recognition Setup

1. Go to Settings and click on the Facial recognition and click on it again then click Continue.
2. Now you can setup the screen lock if you have none already.

3. It will ask you if you have glasses on or not, choose which one and click **Continue**.

Make sure you are holding your phone away with a distance between 8 to 20 inches away, and ensure that your face is positioned perfectly in the circle. The progress bar will start loading if your face is accurately placed until it finishes reading your face for registration. When it is done reading, you can adjust the settings to your desire then click OK.

Additional Tips for Facial Recognition

In case you are facing a problem unlocking your device, these few tips will help you.

You must understand that the best way to register your face is by ensuring that you are not putting anything on like glass, masks, hats, bangs and even beards as this can cause some problems during the process.

Ensure that the area is bright enough for the camera to be effective.

For a great output, ensure that the image is very clear and not blurry.

But I must tell you that face recognition is not well secured compared to PIN, Iris, Pattern or Fingerprint.

If you are an identical twin, your mate can easily unlock your phone.

Setting Up a Screen Lock on Your Phone

Since your phone is known to have all your important information stored in it, you have to make sure it is protected and kept private as much as possible.

To improve the security of your phone, you are advised to set up the screen lock feature on it to prevent intruders from accessing your phone.

Setting up a Basic Screen Lock

The password, PIN or pattern can be activated on your phone to help improve its security thereby preventing intruders from accessing your phone without your permission.

i. Go to Settings.

ii. Swipe up to the Lock screen and tap it.

iii. Tap on the Screen lock type option.

iv. Select your preferred lock type.

v. Then follow the on-screen instructions to set it up.

The available basic screen lock types are as follows:

1. Swipe: With this, you can unlock the screen by swiping.
2. Pattern: Choosing this allows you to create a pattern to unlock your screen by drawing the pattern on the screen.

3. PIN: Choosing this enables you to select a PIN to get your screen unlocked.

4. Password: Choosing this feature allows you to create a password for getting your screen unlocked.

5. None: Choosing this means you will have no screen lock present on your phone.

Setting up Biometric security

The Biometric security enables you to make use of your face, fingerprints or eyes in getting your phone unlocked with the use of the intelligent scan feature.

The intelligent scan makes use of the combination face and iris recognition to perform its function of providing your phone with maximum security.

i. Go to Settings.

ii. Swipe up to the Lock screen and tap it.

iii. Select the Screen lock type.

iv. Select your preferred lock type.

v. Tap the slider which is next to your desired option.

vi. Then follow the onscreen instructions that come up.

The following are the biometric options that are available:

1. Fingerprints: Using this allows you to unlock the screen with fingerprint recognition.

2. Face: Using this allows you to unlock your phone with the use

of facial recognition.

3. Iris: Using this will allow the use of the Iris recognition in unlocking the phone screen.

4. Intelligent scan: Using this combines the face and iris recognition for more effective and maximum security.

Secure Lock settings

When your phone is idle, you can get to set up secure lock settings like the automatic screen lock to provide your phone with extra protection. It is however, important to note that these settings will only appear when a lock screen has already been initiated.

To access this feature, you are required to take the following steps:

i. Go to Settings.

ii. Search for Secure lock settings and select it.

iii. Select the Screen lock settings again

iv. Enter your current lock screen credentials if you are prompted.

v. Tap the sliders to help turn your preferred Secure lock settings on or off.

1. **Make pattern visible:** Whenever the pattern is drawn, choose this option to make sure it appears on your screen.

2. **Lock automatically:** With this, you choose when your phone

screen gets locked once the display turns off automatically.

3. Lock instantly with power key: This helps in setting your phone to lock instantly once you press the power key.

4. Auto factory reset: It is to be noted that after the unsuccessful unlocking of your phone 15 times, it will get automatically reset to its factory settings resulting in all your data including your files and downloaded apps being erased.

5. Lock network and security: This feature helps in keeping the network as well as security-related functions locked while your phone is locked which makes it easier to find and protect it whenever it gets lost or stolen.

6. Show Lockdown option: This feature helps in displaying a Power button which helps in turning off Smart Lock, biometrics unlock as well as notifications on the Lock screen.

Changing the Security Features on your Phone

Your phone is designed to provide you with numerous security features to help in keeping your data secured from intruders.

Setup and use Samsung Pass

First, it is important to note that the Samsung Pass is designed with Samsung internet as well as certain apps. The Samsung Pass is equipped to remember your biometric information which means you don't need to be worried about remembering the passwords and usernames when its need arises for you to get logged into your app or website.

i. Go to Settings.

ii. Swipe to Biometrics and security and touch it.

iii. Scroll to the Samsung Pass and touch it.

iv. Tap on Continue.

v. Select Sign-in and get signed in using your Samsung account credentials.

vi. Select Fingerprints or Irises.

vii. Select Continue.

viii. Then set a preferred secure screen lock.

ix. You will be prompted with on-screen instructions which you will follow.

x. Select Done.

Samsung Pass under Biometrics and security

Once you have been able to turn on the Samsung pass, you can now get the Samsung internet opened and then browse through your desired website. Once you are signed into an account, you will be asked by Samsung pass if you would like to get your account information saved.

The next time you sign in to the site, always tap on remember.

With this feature underway, instead of having to fill in the account information all you will need to do is scan your biometrics.

Setting up and making use of Secured Folder

For maximum security for your personal information, it is advisable to lock your private content and apps in a secure folder which only you have the key.

i. Go to Settings.

ii. Swipe to the Biometrics and Security and select it.

iii. Swipe to the Secure folder and touch it.

Alternatively,

i. Open the Quick Settings panel by swiping down from the top of the screen.

ii. Swipe to the left.

iii. Select the Secure Folder icon.

Securing Folder home page

i. Once your Secure folder is opened, tap on Continue.

ii. Get signed in using your Samsung credentials.

iii. Set up a lock.

iv. Touch next.

v. Once the lock option is set up, your Secure Folder will get created.

Once you have successfully set up your Secure Folder, an app icon for you to access it easily whenever you want to.

i. Go to the Home screen.

ii. To access the Apps screen, swipe up or down.

iii. Tap on the Secure Folder and log in using your credentials.

iv. Tap OK. You will be presented with options to Add apps or Add files to get your folder customized.

v. To get your folder locked, customize your Secure Folder's icon or get access to the Settings, tap on More options.

It is important to note that you will need to get signed in to your Samsung account to make use of the Secure Folder and you will need to set a secure lock on your phone as well.

Using the Secure Startup

To get this feature activated, you will need to set up the screen lock on your phone so that even when your phone falls into the hands of intruders, the Secure startup will be there to protect you.

The secure startup helps in providing your phone with an addiction lock screen which will be required every time to turn on your phone.

i. Go to Settings.

ii. Search for Secure Startup and select it.

iii. Choose the available secure method to set the secure startup. For instance, select Require PIN when the device turns on and tap on Apply.

iv. Input in your current PIN.

v. Select Next to complete set up.

Note that this option does vary based on if a password, pattern, etc is set). It is to be noted that turning on this feature requires that the screen lock on your phone must have been set up. The available options are dependent on the lock screen type or the phone you have. The phone stays unlocked after a reboot if you don't set the lock, automatically feature is not set to immediately or if you don't turn on the Lock Instantly with Power Key setting option.

Chapter 22: Setting up the Ultrasonic fingerprint sensor on your Galaxy phone for use

The biometric is a well-thought technology created for advanced security. You can be rest assured that once your Samsung Galaxy is locked with fingerprint security, it is well secured. Nonetheless, the fingerprint scanner is bound to vary depending on the type of your phone.

Note: Your wireless service provider, phone model or software version will determine your screen and settings variety. To be able to add your fingerprint, you must have a lock screen set up.

Set Up and Use Fingerprint Security

Note: The fingerprint unlock feature will be activated automatically when you add a fingerprint.

Set up and use fingerprint security on Galaxy A Series

Note: The fingerprint unlock feature will be activated automatically when you add a fingerprint.To set it up, follow the instructions below.

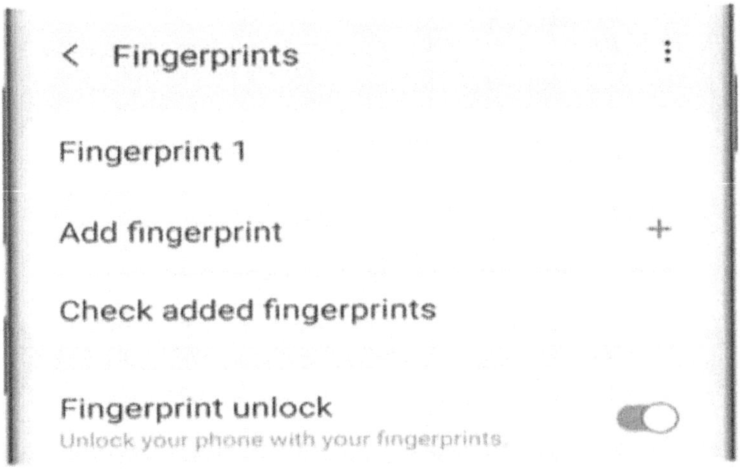

In the Settings, look for Fingerprints and click on it, then input your security details. Then adhere to the instruction that appears on the screen to add your fingerprint.

When you are done, click on the Add and add another. Click on 'Done' when the process is completed.

If you want to unlock your device, click on the screen or Power button and register with the finger you used for the security set-up. This will let the phone unlock automatically once it has recognized your print.

Note: There is a preinstalled plastic screen protector that comes with the GalaxyA Series. That is why you are advised not to install any screen protector as it could affect the fingerprint sensor.

Removing a Fingerprint From Your Phone

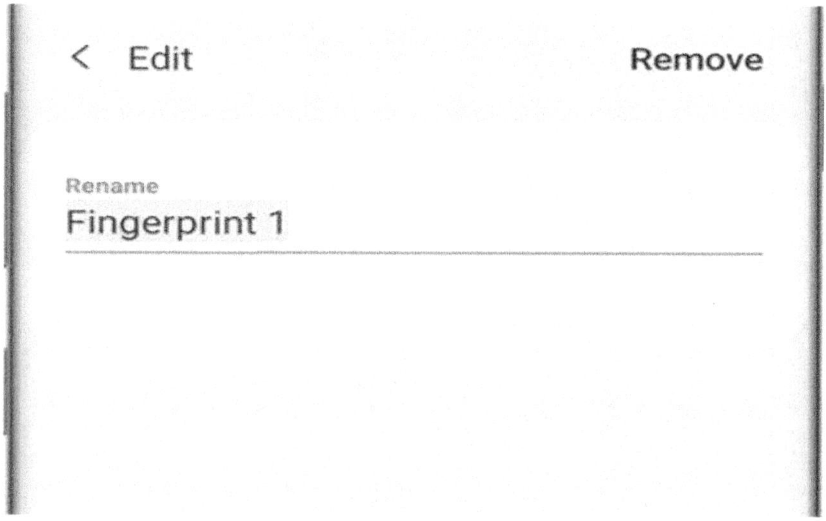

Even though the fingerprint sensors are located in different places, the process of removing the fingerprint is no different with the Pie update on all devices.

In the Settings, look for Fingerprints and click on it, then input your security details. Click the fingerprint you would like to delete for example, Fingerprint 1. Then click Remove. After that click Remove again for confirmation.

Note: The fingerprint security feature will be deactivated when you remove the registered fingerprint.

Turning on Faster recognition on Galaxy A71

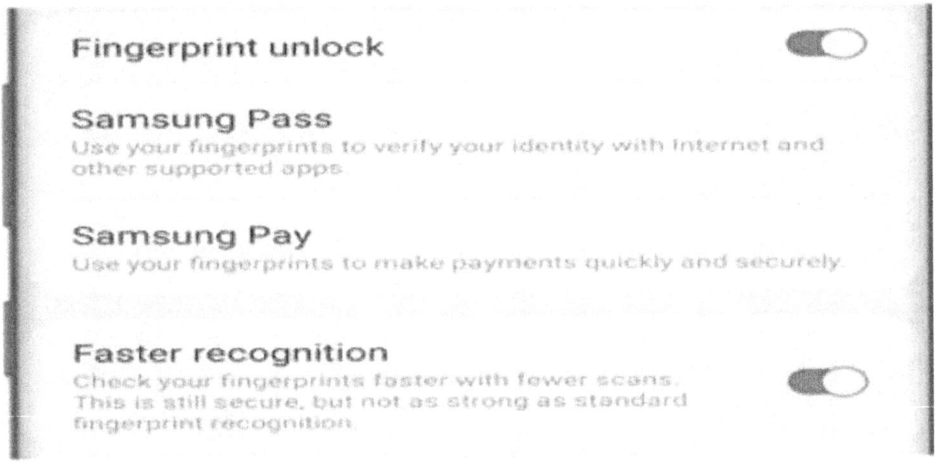

Are you interested in extending the speed of your Galaxy A71? The recognition of your fingerprint will be done faster when you activate the Faster recognition.

1 When a fingerprint is added, go to Settings and click on Biometrics and security soft key.

2. Scroll to fingerprint and click on it and input your lock screen details.

3. Below the screen, you will find the Faster recognition, click on it to get it activated or vice versa.

How to Alter the Secure Lock Settings on Your Device

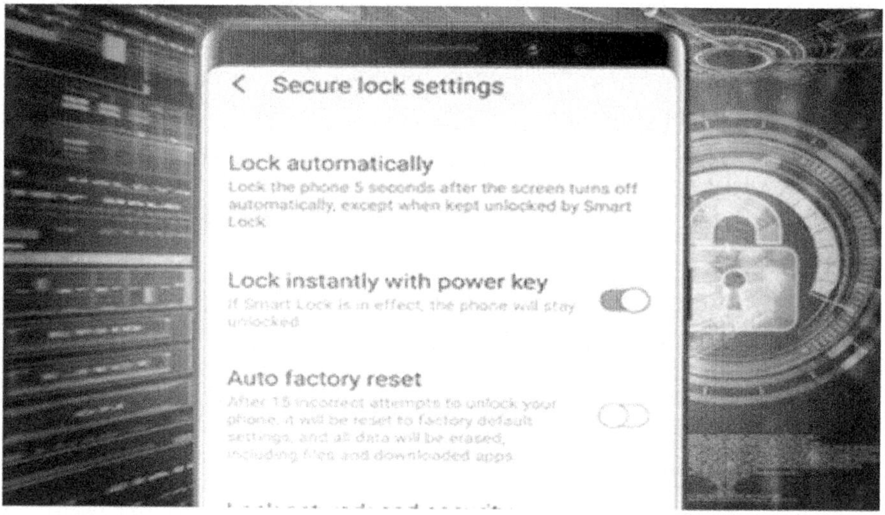

The Secure Lock settings features give you an unwavering security system that put your mind at rest that no one will ever be able to access your phone. It locks the screen automatically according to your preset. On Auto Factory Reset, there are more means for you to secure your device.

Note: Your wireless service provider, phone model or software version will determine your screen and settings variety.

Secure Lock Settings and Features

Important: The secure lock screen can only be available when the lock has been set.

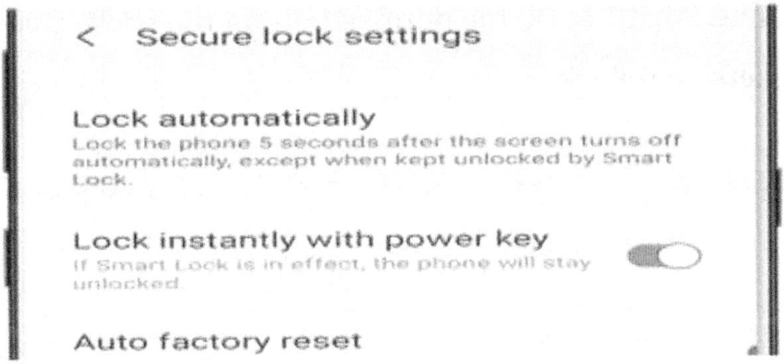

Go to Settings and look for Secure lock settings. Input your secure lock details. Then the below will be available.

Lock automatically: When the screen backlight is off, the screen will lock automatically after a period of a few seconds.

Lock instantly with power Key: The screen will manually be locked when you push the power button.

Note: The phone will be on lock when the Smart Lock is activated.

Auto factory reset: Your phone will reset automatically to factory settings after 15 failed trials to unlock your phone.

Lock network and security: When you phone is on lock, all the security functions and networks will not be accessible. That is to say, you cannot have control over the mobile data and Wi-Fi, which makes it easy to protect your phone even when lost.

Show Lockdown option: This will show the Power key option that disables Smart Lock, the notifications on the Lock screen and biometrics unlock.

Chapter 23: Manage Recent Apps on Your Phone

The Galaxy A71 device was made to perform different tasks at once. This means it can easily switch between apps. For instance, if you are playing a game and you get a message alert, you can just check the message and afterwards go back to the game.

To do this:

- Simply use the navigation bar and
- Click on Recent. There, you will find all recently opened apps.

Note: The type of internet service provider or phone model will determine the screen display and Settings.

What You Can Do In Recent Apps

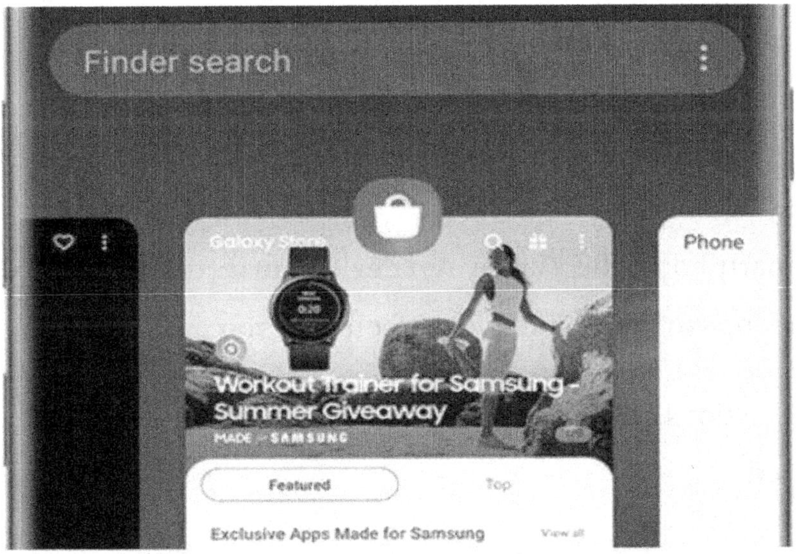

You can manage the apps you find on your recent apps.

Open 'Recents' from the left side of the home screen.

- Press the app you want to open from the list of apps provided.
- Flick over to the left or right to see more apps
- Flick an app up if you want to close it
- Press 'Close all' if you want all the apps closed.

Turn On Suggested Apps

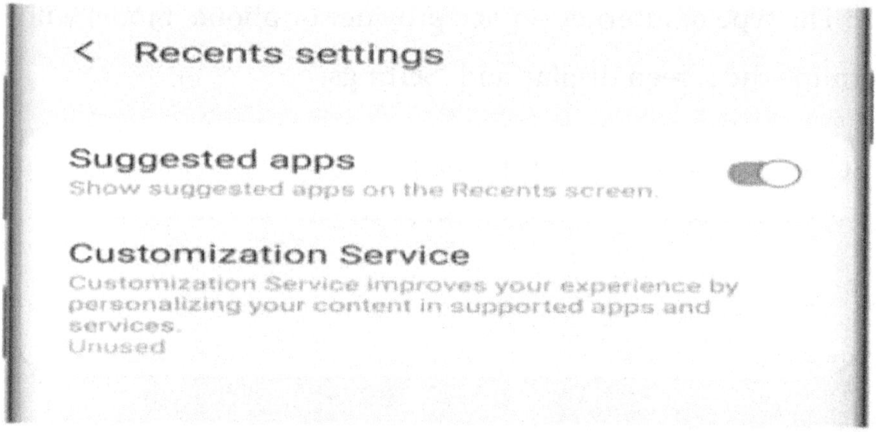

Being a smartphone, the Galaxy A71 device can introduce some important apps to you which depend on how you use it.

These suggestions can be seen at the bottom of the recent apps you opened.

- Simply open recent apps from your home screen.
- Press 'More Options'
- Press 'Settings'
- Press the pointer beside Suggested apps to switch it on

Capture the Flavor of Your Dish with Food Mode

With Food Mode, foodies and food photographers get a filter which automatically adjusts the colors to give your food, eye popping delicious looks.

Capture That Beautiful Dish In Food Mode

Why hide that beautiful dish from the world? Let it bask in the spotlight. Switch your camera app to Food mode, so you can do it some aesthetic justice.

Enter Camera app and switch to food mode. Keep your dish in focus and use the wheel icon to adjust the color temperature till you are satisfied with the result.

You can use the Radial Blur icon on the left of the wheel to keep the background out of focus. When you are satisfied with that awesome result, capture the moment.

Note: Food mode may not be in default. Enter the Camera app then go to settings. Find and click Camera Modes then click Edit modes. Select Food and then return camera homepage.

Chapter 24: Create an Emoji of Yourself

If you enjoy sending selfies and emojis to your loved ones, this is the right phone for you. You can make an emoji of yourself and even send it in text messages.

Create and Edit Your Emoji

To see the emoji version of yourself,

1. Open your camera app and click on the **AR Emoji.**

2. Click Create **My Emoji** then allow.

3. Follow the instructions on your screen then capture the image when you are satisfied.

4. Choose your preferred body type then click next.

Congratulations on meeting your very own emoji.

If you are not satisfied with the result, you can edit the details before saving your emoji.

You will find the edit icon in the upper right corner of your screen, you can choose different categories and make changes to the features till you are satisfied. Click the Check mark to save the result. Enjoy your emoji.

Note: For some phones, you can't edit the details of your emoji after saving it. Once you save the final version you won't be able to edit it.

Personalize Text Messages with Your Emoji

The whole point of having your own emoji is to show it off.

This phone lets you customize your messages with them and it's pretty easy to accomplish.

Enter the messages app and create a new message. Click on the Enter message field to show the onscreen keyboard. Click on the square smiley face icon, this is called the stickers icon.

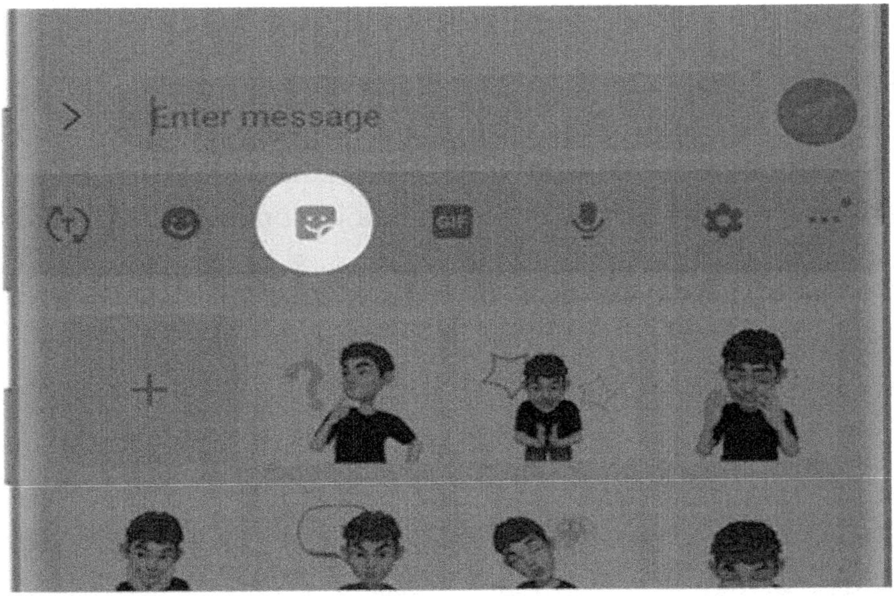

There you will find different GIFS of your avatar. Choose the desired emoji and send.

Delete Emojis and Emoji Stickers

Do you want to change your emoji's avatar? Are you bored or looking for something to express the new you? Don't panic you can delete your old emoji.

Enter your camera app and click on AR Emoji icon. Here you will find your emojis, Tap and hold on the emoji you want to remove until delete icon appears. Click delete and confirm when asked if you want to delete the emoji's sticker and Voila.

Note: This option is only available on select phones.

The remove my emoji icon also called the stop icon, does not delete the emoji.

It only changes the camera mode back to regular live feed from AR Emoji mode. The stickers of the emojis you deleted may still be in your Gallery. They take up your storage space especially if you have made lots of emojis. Try to delete these stickers from your gallery to save space.

To do this, enter your gallery app and click on albums. Click on the Emojis folder and search for the undesirable stickers. Select and delete them.

The emojis have new designs and more customization features on new phone. You can make an emoji based on your age, you can make it an adult or a child. You can choose a desired gender. Options include male, female and if you want to, you can make it gender neutral or androgynous. You can name your emoji and edit it anytime you feel like. You have the option to download new outfits from Galaxy Apps.

New features include the ability to react with your emoji in the Emoji home screen and custom stickers. After creating an emoji, click AR Emoji and the click studio to visit the My Emoji home page. This is where all the customization features are available for your pleasure.

Chapter 25: Cameras on Galaxy A Series

Galaxy phone cameras have increased with every new release. Therefore it is not surprising that the Galaxy A71's camera is incredible. New modes added to the camera include Live Focus Video, Night Mode among others. By the way, your favorite camera modes like Super Slow-mo are still available.

Live Focus Video

This image showing a girl playing with water is from a live focus video. Upgrades to the video recording features are mind blowing. The live Focus modes have now been applied to videos and focus effects have also been added to make your subjects, be it people, animals or objects stand out. These functions apply to both the front and rear camera.

To use the live focus video, enter the camera app and swipe to Live Focus video. Choose your desired effect but know that these filters may not apply until the camera detects a face.

Basic blur: blurs the background to keep your subject in focus, giving the video a film look.

Color tone: It reduces the tone of the background to grey, ensuring that the subject is getting all the attention. Adjust the level of grey with the slider.

Big circle: make your subject the focus by altering the background with garbled light orbs. This gives your image an ethereal look.

Glitch: Gives videos a shaky and amateurish look, like a horror movie.

Super Steady

When you make a video on your phone, the result is a shaky hand held like footage. The super steady option on your Galaxy A71 helps you make stable, super smooth videos.

1. Enter your camera app and choose the rear camera.

2. Choose video or hyperlapse option and activate the super steady icon at the top of your screen.

3. Open the Camera app, and navigate to the rear camera.

4. Select either Video or Hyperlapse, and then tap the hand icon at the top of the screen.

This is the super steady icon and it glows yellow when activated. The super steady option works wonders on wide angle and ultra wide angle lens.

Note that super steady only works on video and hyperlapse mode with the rear camera.

Night mode

The night mode feature on the Galaxy A71 adapts your phone camera to low light surroundings for brighter pictures. This feature ensures the pictures you take at night are a lot better than that of earlier models.

To apply it, enter the camera app and choose night mode. When you take your pictures in this mode, the Galaxy A71 captures the images individually and stitches it together into one bright and wonderful shot.

Zoom In Mic

Unlike older models that can't record conversations from afar. The Galaxy A71 has a zoom mic. This feature increases the volume of the audio recording, as you zoom in on the source of the audio, while minimizing the background noise.

Although it should be by default, ensure this function is activated. To confirm

1. Open the camera and choose the rear camera option.

2. Select video and then click on settings.

3. Click on advanced recording options, there is a switch beside the zoom in mic option. Make sure it is on.

The Galaxy A71 has three microphones instead of two, this ensures better and clearer sound recordings.

Note: The Zoom-in mic option can fade out, depending on the camera mode.

Chapter 26: Transfer Media Files on Your Samsung

Phone

If you have a large amount of pictures, music or videos taking up your phone memory, there are lots of options for you to transfer your media flies.

You can move them to a microSD card, the cloud, a PC or the smart switch.

Use a microSD card

When you need to create space on your Galaxy A71 you can easily make use of a microSD card. The device care app is also an option to clean up your phone.

1. Insert the microSD card into your phone,

2. Then in the Samsung folder app, click the My Files folder.

3. Locate the files you intend to move and then touch and hold them until options emerge.

3. After marking the files, click either the move or copy option according to your intent. They are available at the bottom of your screen then return to the My Files homepage.

4. Click on the SD card then click move.

You can't copy or move files like this in any app other than your My Files app.

5. Also you must insert a microSD card to have this option.

Note: Your phone must support microSD cards and you won't be able to move encrypted music onto your SD card.

Transfer to A PC or Laptop

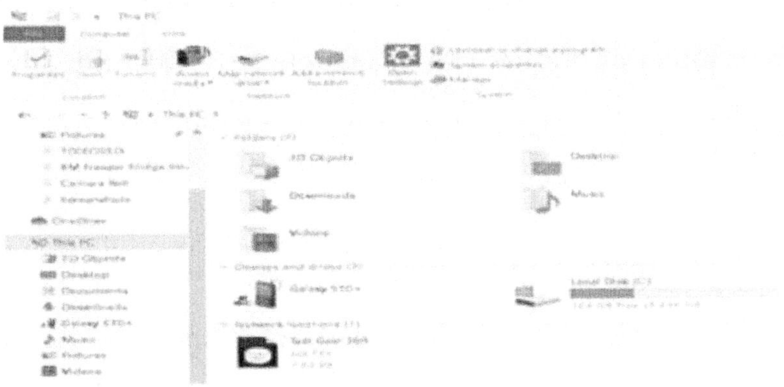

You also want to transfer your files to your Computer by connecting your phone to the computer or tablet with a USB cable. After the connection is established, use the File Explorer on your Pc to move the files. If you want to transfer your files wirelessly the Samsung Flow is an available.

Use a Cloud

The Samsung Cloud is a secure database to store your files. Sign into your Samsung account to use this option. It allows you to download your pictures, music and video files anytime you need them. If the Samsung cloud is not available to you, you can store your files on other cloud services.
Your options include One Drive, iCloud, Dropbox and Verizon cloud.

Use Smart Switch

The Smart Switch is another option to transfer your media files from your tablet or phone. This is a wireless transfer app. It is an app that lets you transfer your files from one tablet or phone to another and even to a computer.

Use the Trash feature in Gallery

The gallery app has a recycle bin called the trash folder. All your deleted media files are stored here for easy recovery. After 15 days, these files will be deleted permanently.

To delete a file enter the Gallery app and select the image you want to delete.

1. Click the trash icon, and then click move to trash to confirm delete.

2. To view the media files in your Trash, return to the gallery homepage.

3. Click on **More option** and then select Trash. This opens your trash folder.

4. If you want to delete the files permanently, just select empty trash and they will be gone from your phone.

5. You can use the Edit option to delete only selected file(s) or restore them to your Gallery.

Chapter 27: Using AI Assistants on Your Phone

Many AI assistants can be used on your phone amongst which we have Bixby and Google assistant. With the availability of Alexa, the Samsung internet can also be set as a phone assistant as well. The Bixby has been specifically designed for use on Galaxy phones along with its special features. Nevertheless, you can have the assistant on your phone changed from its default if you so desire.

Choosing Bixby As Your Default Assistant

Bixby is the only AI assistant designed among others to access services that are exclusive to Samsung. They can be used on your phone to perform some important functions such as:

1. You can get to control your devices in SmartThings.
2. They provide you with the ability to make purchases with the use of Samsung pay.
3. You can stay tracked with your schedule with the help of the Bixby reminders.
4. Bixby can get connected with your Spotify account thereby making your search for songs easier and faster.
5. With the presence of the Bixby vision, you can browse and get your shopping done.

7. They are made of amazing key functions.
8. Your daily routine gets easily adapted to by the Bixby Home.
In general, other AI assistants are made up of features that are similar to those of the Bixby but they are not integrated in the same way as Bixby with your phone.

How to Change Your Default Phone Assistant (One UI)

Bixby always makes itself available for use even with the presence of other assistants. It doesn't, however, stop you from making other assistants as your default if the need arises as an addition to Bixby itself.
i. Go to the Settings.
ii. Search for "Device assistance app" and Tap it.
iii.Tap "Device assistance app" again and you will be prompted with a list of available assistants.
iv. Select your preferred choice.
v. Tap OK.
It is to be noted that whenever you press the Bixby key, it will stay active and always wake up. You will also have the opportunity to always get your Bixby button remapped.
Nevertheless, it is important to note that this feature is referred to as a Phone assistance app on some phones.

How to Change Your Default Phone Assistance (Oreo OS)

In a situation where you want to try the Google assistance out for a while, it can be made your default AI assistance.

i. Go to the Settings.

ii. Search for "Device assistance app" and Tap it.

iii. Tap "Device assistance app" again and you will be prompted with a list of available assistants.

iv. For Galaxy phones, there won't be Bixby on the list since it will always be the main voice assistant.

v. Select your preferred option.

As pointed out earlier, note that whenever you press the Bixby key, it will stay active and always wake up and similarly, your Bixby button can be remapped as well if you want to. Also, the feature is referred to as the Phone assistance app.

Chapter 28: Using the Night Mode on Your Phone

The new Night Mode feature from Pie update provides you with the opportunity to make use of your phone at night while lying on your bed without causing any hurt to your eyes with its brightness. This Night Mode feature helps in making the theme of your phone darker for comfortable use during the night time.

Turning On the Night Mode Manually

Once the Night Mode feature is turned on, the menu of your phone settings as well as preloaded Samsung apps will make use of the darker theme.

Nevertheless, third-party apps will stay the same.

You are required to take the following steps to help activate your Night Mode:

i. Open the Quick Settings panel by swiping down from the top of the phone screen.

ii. Then swipe to the left of the screen.

iii. Tap the Night Mode icon which will cause it to glow blue to indicate that it is on.

iv. Tap on the icon again to turn off the Night Mode.

Setting a Schedule for Night Mode

There are times you will be so busy and carried away such that you won't know its dark outside and you have forgotten to turn on the Night Mode feature on your phone. It is therefore advisable to have your Night Mode set to automatically turn on or off so you won't have a problem keeping track with time. You are required to take the following steps to do this:

i. Go to Settings.

ii. Search for the Night Mode and select it.

iii. Tap on the Night Mode again.

iv. Tap on the slider which is next to the "Turn on as scheduled" option.

v. To turn your Night Mode on and off in the evening and morning respectively, select Sunset to sunrise.

To make use of a Custom schedule in setting your schedule for Night mode usage, take the following steps:

i. Select the Start time.

ii. Select your preferred time.

iii. Tap on Done.

iv. Then touch the End time and choose a time.

v. Tap on Done.

It is to be noted that even if scheduled time is set for Night Mode, it can still be turned on or off manually whenever you want to.

Chapter 29: Managing Apps on Your Galaxy Phone

The Galaxy phones are designed with apps and you can also get to download all your preferred apps. You are also allowed to get apps which you don't want or don't use anymore uninstalled or disabled. Once you have a good understanding of how to get your apps managed, you will have the chance to maximize the potential of your phone.

Downloading App

The Google Play store, as well as the Galaxy Store, is the two app stores that are available on your phone.

Go to the apps screen or go into the Google folder to get access to the Google Play Store while the Galaxy store is available in the Samsung folder or check the Apps screen for it.

With the use of the sort and search options, pick up an app store and check for available apps.

As soon as you come across the apps you like to download, tap on it and finally tap INSTALL.

It is important to note that only apps that are supported by your phone will be available for download.

Uninstalling Apps

i. Go to the Home screen.

ii. To access your Apps, you should swipe up or down.

iii. Touch and hold the app you like to uninstall.

iv. Select Uninstall.

It is however, important to note that you can't uninstall or disable some preloaded apps.

Disabling Preloaded Apps

In a situation where your phone is made up of preloaded apps that you don't use, you can't get them completely uninstalled but on the other hand, they can get disabled.

i. Go to the Home screen.

ii. To access your Apps, you should swipe up or down.

iii. Touch and hold the app you like to disable.

iv. Select Disable.

Reinstalling Paid Apps

In a situation where you have already gotten an app purchased and you don't want to go through the process again. All you will need to do is get it reinstalled. If it was purchased using the Play store, do the following:

i. Open the Play Store.

ii. Tap on Menu.

iii. Select My apps & games.

iv. Select LIBRARY.

v. Find and INSTALL your preferred app.

If the app was originally installed with the use Galaxy Store, you will need to take the following steps:

i. Open the Galaxy Store.

ii. Tap on More Options.

iii. Tap My page.

iv. Select All.

v. Go to the Apps tab to search for the missing app.

vi. Tap on the Download icon option.

Managing Your App Permissions

For some information to be accessed on some apps, such apps may require you to ask for permission for certain features to be used. An example of such apps asking for permission includes the Email app asking for permission to see your Contacts for it to have access to import the contact info from your friends.

In a situation where you have granted permission accidentally, the settings can always be changed by taking the following steps:

i. Go to the Settings.

ii. Select Apps.

iii. Select your preferred app.

iv. Swipe to Permissions and tap on it.

v. For the permission you want to deny, tap on the switch that is next to it.

Protecting Your Phone from Viruses and Malware

You need to take note of some fake ads such as "WARNING: YOUR PHONE IS INFECTED!" This kind of ads are entirely untrue because all Galaxy and Play Store apps are scanned first before they get downloaded as a result it is very unusual for your phone to get infected with viruses and malware.

Nevertheless, harmful software can try to get downloaded onto your phone via sneaky ads or emails which is why you need to get your phone protected. There are laid down procedures to help in guiding against this occurrence.

App Permissions

The App permission helps in protecting your phone from damage. Whenever permission comes up, you need to be sure before granting access. Once in doubt of an app, it is advisable to avoid granting access.

Once the installation of a new app is complete, you will be prompted with a permission message which you will need to allow or deny.

You will have to choose between the **deny and allow** options depending on the app and the user.

Browser Settings

Apart from untrustworthy sites, you can still be prompted with unsafe or annoying ads at times from even the known trustworthy sites. To avoid being conned by scammers, you can stop all the unsafe pop-ups with the use of the options on your Samsung internet.

i. Go to the Samsung folder in Apps.

ii. Tap on Samsung Internet.

iii. Tap Menu.

iv. Tap on Ad Blockers which will prompt you with a list of recommended content and ad blockers available for your browser.

v. Select an app.

vi. Select Install to get the content blocker.

Note that in a situation where you are making use of a different browser app, always remember to get its security settings adjusted.

Removing an Infected App

In a situation where you feel your phone has gotten infected, you will need to take the following steps to remove viruses:

i. Turn off the phone power completely and put it back on.

ii. Press and hold down the Volume Down key once the Samsung logo is displayed.

iii. At the bottom left corner of the screen, the "Safe mode" will appear if the previous steps are carried out correctly.

iv. In a situation where the Safe mode doesn't come up, you will need to repeat the previous steps.

Take note that once the phone is in the Safe mode, all third-party apps will get disabled which makes a great way of finding out the cause of the problem.

The Galaxy Phone Home Screen with Safe Mode Displayed

You will need to take the following steps to help you in removing potentially infected apps:

i. Go to the Settings.

ii. Select Apps.

iii. Select the preferred app.

iv. Tap on Uninstall.

v. Restart your phone and reboot normally to exit the Safe Mode.

How to Make Sure Your Phone Is Safe

To help keep your phone free from viruses and you want to be completely sure your phone is safe, it is advisable to take note of the following tips:

1. Avoid download apps apart from the ones in the Play Store or Galaxy Apps.

Samsung not support the use of anti-virus products from third parties, so it is advisable to avoid the use of antivirus apps.

2. Make sure that your phone is updated with the latest software as well as the operating system.

Chapter 30: Device Care Options on your Phone

The Device care feature helps you to keep track of your phone's performance.

With this feature, you will get an overview of your battery, RAM, storage as well as Security.

The feature also enables you to automatically get your phone optimized for faster and much better performance.

Quick Optimization

The performance of your phone is improved by the quick optimization feature by helping you to identify apps that's exhausting lots of battery power while also helping in the clearing of unwanted items from the memory of your phone, deleting unnecessary files and closing apps that are running in the background.

Follow the instructions below to help you in optimizing your phone:

i. Go to Settings.

ii. Swipe to Device care and tap it.

iii. Tap Optimize now

iv. Select Done.

Optimizing Battery and Using the Power Saving Modes

The use of the battery in phones is its lifeline and there are different ways by which you can get your battery settings changed to optimize the battery life of your phone.

i. Go to Settings.

ii. Swipe to the Device care and tap it.

iii. Select Battery which will show you a brief overview of your battery usage.

You can make use of the different power saving modes to get your battery power saved.

Select the Power mode and choose from the following options stated below:

1. High performance: This provides maximum screen brightness as well as resolution that makes it drain more of your battery power.

2. Optimized: This provides a recommended balance of performance as well as battery life.

3. Medium power saving: This provides extended battery life by limiting some functions.

4. Maximum power saving: This will help you in saving as much battery life as possible.

Alternatively, you can gain access to your battery options by tapping on the More Options and then selecting Setting. To get a detailed report about your usage record, tap on the Battery usage.

Clean Storage

In a situation where your phone is being over-occupied with unnecessary residual files, all you need to do is get your storage cleared.

i. Go to Settings.

ii. Swipe to Device care and tap it.

iii. Select Storage.

iv. Tap on CLEAN NOW to get storage space freed.

Once these steps are followed correctly, all unwanted data will be cleared.

Clean Memory

Once you discover that your phone is slow and getting glitch, it's advisable to get the memory of your phone cleaned up to get its performance enhanced by taking the following steps:

i. Go to Setting.

ii. Swipe to Device care and tap on it.

iii. Tap Memory.

iv. Tap on CLEAN NOW to get your memory freed up by stopping all apps running in the background.

Rearrange Apps and Widgets

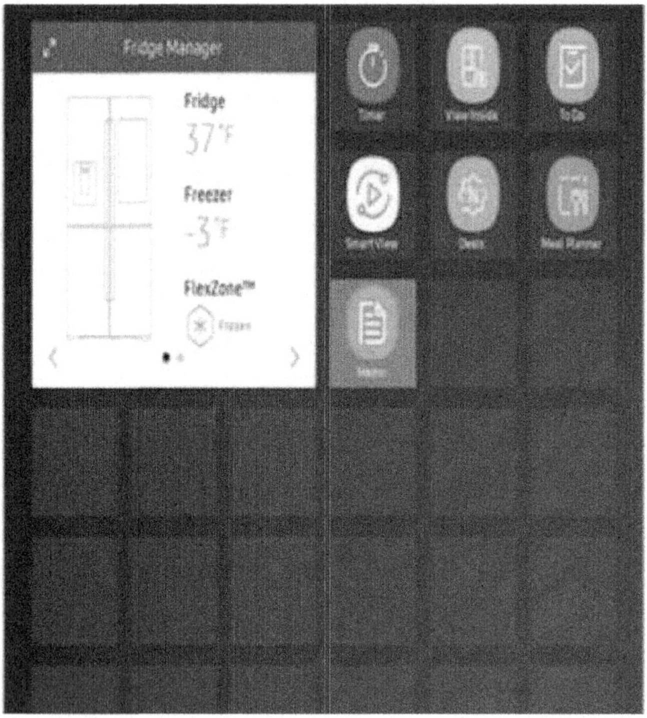

You can drag and move app to any desired place on the Family Hub Home screen, the same way you do your smartphone. Place the widgets and apps by yourself so that you will know where to look for an app when needed.

1. Press and hold on any app or widget you would like to move to a particular location and drag it there and then, you release your finger.
2. Then, you have successfully transferred an app to your preferred destination.

Apps icons come in similar sizes but widgets, resize the widget to fit to where you want it to be.

Removing An App, Widget, Or Home Screen

Apps or widgets can be removed for the home screen except the Apps icon. You can as well get rid of the whole pages on the Home screen except the first one.

You can get rid of an unwanted app or widget by pressing and holding the app and then click Delete.

Press and hold on a free space on the home screen and click on the Delete soft key seen at the top right corner of the screen.

Customizing Your Wallpaper

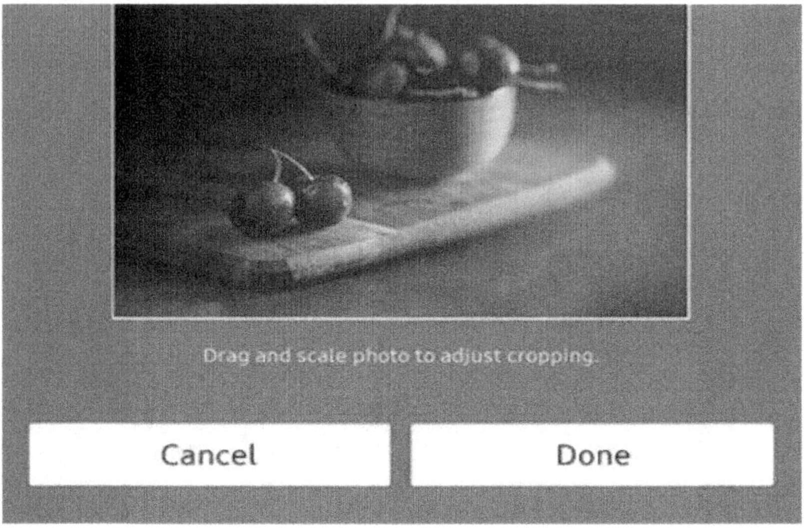

Family Hub has already gained the attention in your kitchen, thus, you can make it more beautiful and appealing with a customized wallpaper.

1.Touch Settings, touch Display, and press Home screen wallpaper.

2. Click the Wallpaper to check for some preloaded wallpapers or you can press Photo to choose any image of your choice from your gallery. And set it as wallpaper. Click Done when you are done.

Chapter 31: Find Your Samsung Galaxy Devices

Whether you have lost tablet, phone, earpiece or even watch, do not fret, the **Find My Mobile** is a feature that lets you locate or unlock and lock and even erase your data. Even the Samsung Pay payment info can all be cleared.

Find your phone or tablet using Find My Mobile

Note: Ensure that all your Find Mobile requirements has been put in place just in case it is lost.

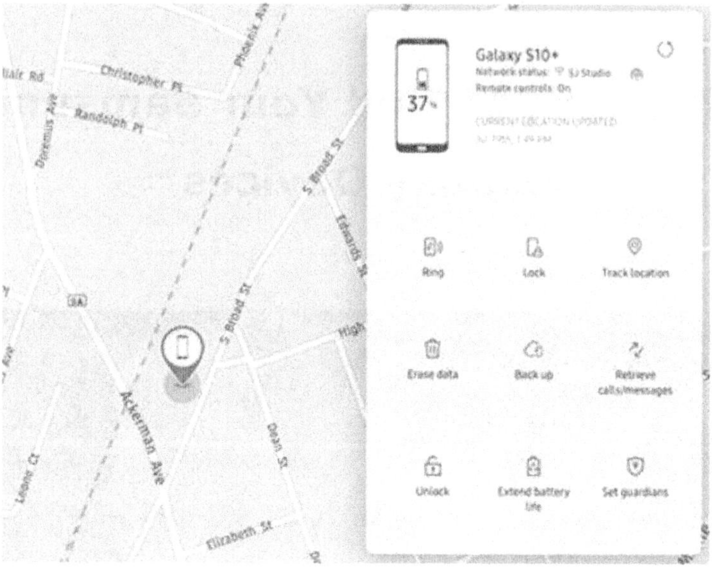

Find MY Mobile can give you a precise place of where your lost phone or tablet is.

Visit Find My Mobile website preferably on PC and input your Samsung account details, then solve the I am not a robot captcha and log in.

You will see the device that is linked to your Samsung account. When a device is on and is connected to a network it will show under the **My Devices.** Find My Mobile interface and device location map.

Here are some of the features you will see in the Find My Mobile Interface

Ring: Ensure that you activate the Ring option in case you don't know where you misplace your phone.

Lock and Unlock: If you are concerned about your data, you can secure it by remotely lock your device. Doing this will enable the Lock Screen and expel the card that you have with Samsung Pay. You can also remotely unlock your device in case you forget you lock method.

Note: If unlock or lock process failed, there will be a red light that will blink on to top of the phone.

Track location: This lets you track your phone location every 15 minutes.

Caution: Be careful in the retrieving process, it is dangerous to go to retrieve your stolen phone on your own, rather send the authorities to get it for you.

Erase data remotely: Do not let the burglar have access to your information. Your Data will be cleared including Samsung Pay.

Note: You will be required to sign in back with your Google account to continue using your phone.

Back up: Samsung Cloud will receive your data, so ensure that you backup before erasing them.

Extend battery life: Use these features for your device to stay on for a longer time.

Retrieve calls/messages: This lets you view the last 50 calls on your device

Set Guardians: You can customize guardians that will find your device remotely.

Enable the Emergency Mode or ring the device.

Note: If after all this, you cannot locate your device, the carrier could help in the search or suspend your phone. Reach out to your customers support for more info.

Find Your Phone Using Your Smart Watch

In case you left your phone somewhere accidentally, Samsung Smart Watch can help you locate it.

1. Click on the Home button on your Smart Watch, and tap Find My Phone.
2. Press Start to commence the search. Your phone will ring and then you will be able to get its location. After you have found it, you can click on the dismiss icon to end the search.

Find Your Phone And Tablet's IMEI, Mmodel Number, Or Serial Number

IMEI also known as MEID, serial number are numbers that comes with your phone model as a form of identification and originality. They differ for every device. They help you recognize your phone in case it gets stolen. And there is a special promotion from Samsung which you can only claim using you IMEI or serial number.

Note: Your wireless service provider, phone model or software version will determine your screen and settings variety.

Difference Between Serial Number And IMEI/MEID

The serial number is a number given by the phone maker like Samsung for the user to be able to identify their device, be it TV, watch, tablet and more.

This gives the manufacturers to be to record their products while IMEI or MEID serves the same purpose but it is used on communication technology devices. MEID or IMEI are different from serial number but your phone or table must have it depending on their carrier. They are not only used by the makers but also at international level.

A model number is a unique number that comes with your phone model or tablet. Mode number varies based on the released year, the carrier, the color and other factors.

How to Find Your IMEI, Serial Number and Model Number

It is quite an easy task to find your serial number, model number and IMEI. To view this: Go to Settings, and stroll down to About phone or tablet.

Screen Displaying Info On About Phone

The back of your device: If your device's battery is a removable one, you can check below the battery for the serial number and IMEI.

The packaging: There may be the IMEI and serial number on the box.Your carrier's website: Your carrier website may list some details about your phone that may include the IMEI and serial number.

Bill of sale: Check your sales receipt for phone's IMEI and serial number.

Samsung Members: If you have a Samsung Member account, you can find it below your profile page.

Dial *#06# on your device to show the IMEI and serial number.

Note: This will only work if your carrier is support it.

Chapter 32: Advanced features

Dual Messenger

The dual messenger creates a platform where you can make use of two separate accounts for the same messenger app once the second app has been installed. To install the second app, you are required to take the following steps:

i. Go to the Settings screen

ii.Tap “Advanced features.”

iii.Select the “Dual Messenger,” which results in the display of supported apps

iv.To install the second app, tap the app switch

v. Once you have been able to follow these steps correctly, the second app will get installed with its icon displayed along with it.

Every time you get new notifications from the second app installed, you will have the notification displayed along with it to help to distinguish them from the notifications of the first app.

NOTE

- The dual messenger feature availability depends on the app.
- Features available for use may be limited to the second app.

How to uninstall a second app

i. Go to the Settings screen

ii. Tap "Advanced features."

iii. Select "Dual Messenger."

iv. To uninstall, tap the switch of the app and select the Disable or Uninstall options.

Once these steps are correctly followed, all data that are related to the second app will be removed and once the first app is uninstalled, the second app will be deleted.

DEVICE CARE

With the help of the device care feature, you will get access to view important information about your device such as:

- Battery status
- Storage
- Memory
- **System security**

Apart from getting to view important information about features listed above, you will also get to optimize your device automatically with a tap of your finger.

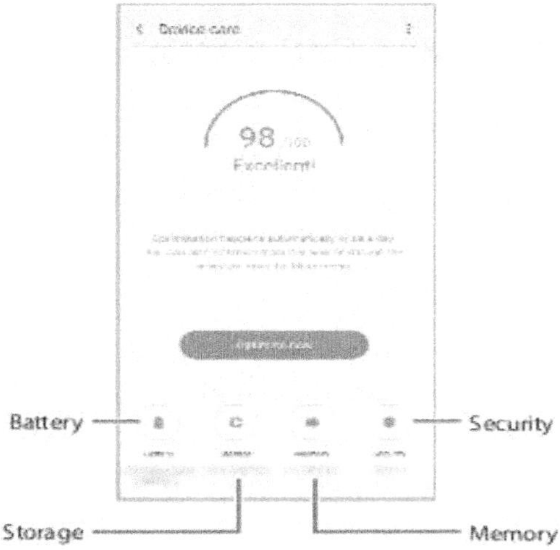

How to optimize your device

i. Go to the Settings screen

ii. Tap Device care

iii. Select "Optimize now" option

Your device performance can be improved by making use of the quick optimization feature via the following actions:

- Getting some memory cleared
- The management of battery drains
- Scanning for crashed apps as well as malware
- Getting unnecessary files deleted and closing running background apps.

Making Use of the Auto-Optimization Feature

If your device is not in use, you can set it to automatically perform optimization by taking the following steps:

i. Tap Auto optimization

ii. Tap on the switch to get it activated

iii. Tap "time," to select what time you want auto-optimization to be carried out.

Battery

With this feature, you can easily check for your remaining battery power as well as device use time. When your device is at low battery levels, the power saving features can be activated to help in saving your battery power.

i. Go to the Settings screen

ii. Tap Device care

iii. Tap Battery

The usage time left feature: This helps in showing the time left before your battery power runs out. It is important to note that the Usage time left is dependent on your device settings as well as operating conditions.

While making use of the power saving mode, notifications may not be received from some apps. The device battery can be set to charge more quickly by going to your device settings and selecting the Fast cable charging switch to get it activated.

Power Mode

It is advisable to select a power mode based on what works for your device.

Select the Power mode and you will be prompted from the following options which you can choose from:

High performance: Selecting this makes your device's screen as bright as possible and provides it with a maximum resolution. This may, however, cause your battery to drain quickly.

Optimized: Selecting this provides a balance between the device's performance and the battery usage

Medium power saving: Selecting this helps in activating the power saving mode to have the battery's usage time extended.

Maximum power saving: Selecting this helps in the reduction in battery reduction of your device by getting the night mode activated and limiting the apps and features that are available.

Note that network connections with the exclusion of only the mobile network will be deactivated.

Adaptive Power Saving

Depending on your usage patterns as well as the battery power you have left, you can set your device to get power-saving mode activated automatically. To do this, do take the following steps:

i. Tap "Power mode"
ii. Tap the Adaptive power saving switch which activates it.

The power-saving mode will be activated by your devices based on usage to get the battery's usage time extended.

Managing Battery Usage for Each App

If you are not making use of some apps, you can stop them from running in the background to help in saving battery power. All you need to do is getting apps selected from the apps list and tap on the **Put app to sleep** switch features to get the feature activated.

Storage

With this feature, you will get to know the status of your storage used as well as available memory.

i. Go to the Settings screen

ii. Tap Device care

iii. Tap Storag

NOTE:

- The available capacity of the internal memory is less when compared to that of the specified capacity simply because the operating system and default apps occupy part of the memory. However, when your device is updated the capacity available will change.

• The Specification section of your device which can be found on the Samsung website provides you with the available capacity of the internal memory.

Managing the Memory

i. Tap "Clean now," to get residual files such as cache deleted.

ii. Under the user data, select a category to help in deleting files or uninstall apps that you are not using anymore.

iii. Select to tick items to be deleted and finally tap on the "Delete" option.

Memory

i. Go to the Settings screen

ii. Tap Device care

iii. Tap Memory

iv. Go to the apps list, tick apps and select the Clean now option to help in speeding up your device by stopping running background apps.

Security

With this feature, you will get access to the security status of your device as well as get to scan your device for malware.

i. Go to the Settings screen

ii. Tap Device care

iii. Tap Security

iv. Select Scan phone

Apps

This feature helps you in managing your device's apps as well as getting their settings changed. The following functions can also be carried out using this feature:

- Viewing the app's usage information
- Changing the notification or permission settings
- Uninstalling or disabling unnecessary apps.

All you need to d here is go to your Settings screen and select the Apps option.

General Management

With this feature, you can get to customize the system settings of your device or reset the device.

i. Go to the Settings screen

ii. Tap General management

iii. Then you will be prompted with the following options:

- Language and input: With this option, you will get to select your device languages and change settings like your keyboards and voice input types. Options vary based on your choice of language.

- Date and time: With this option, you can get to view and change settings for controlling the way time and date is being displayed by your device. It is important to note that time and date will reset if the battery remains fully discharged.

- Contact us: Here you will get to ask questions or view frequently asked questions.

- Reset: Here you can reset your device's settings as well as get to perform a factory data reset. Settings that will be affected by your reset include all your settings or only network settings or the accessibility settings. The device can also be set to restart at a preset time for device optimization.

How to Add Device Languages

You can add Languages to your device by taking the following steps:

i. Go to the Settings screen

ii. Select "General management"

iii. Tap "Language and input"

iv. Tap "Language"

v. Select "Add language"

vi. Tap on "All languages" to view all languages that can be added.

vii. Select a language to get added

viii. Tap "Set as default" to set the chosen language as your default language or otherwise tap "Keep current" to help in keeping the current language setting.

Once these steps are followed correctly, the language that was selected will be included in your language list. In a situation where the default language is changed, the chosen language will be included to the top of the list.

For the default language to be changed from your language list:

i. Drag next to a language and get it moved to the top of the list

ii. Tap "Apply"

If the default language is not supported by an app, the next supported language available on the list will be used.

Software Update

The software of your device can be updated using the firmware over-the-air (FOTA) service and this software updates can also be scheduled as well.

i. Go to the Settings screen

ii. Select "Software update"

Every time emergency software updates are available for your device's security as well as blocking new types of security threats, they will get installed automatically without the need for your agreement.

Download and install: This option enables you to check for updates and get them installed manually.

Auto download over Wi-Fi: This option allows your device to get updates downloaded automatically once connected to a Wi-Fi network.

Last update: This option enables you to view the last software update's information.

Chapter 33: Key settings you need to change in your GalaxyA71

Make use of the volume key for media and turn on Dolby atmos.

i. Go to Settings

ii. Select "Sounds and vibration menu"

iii. Toggle on Use Volume keys for media so that the volume rocker will prioritize getting media volume controlled by default.

iv. Toggle on the Dolby Atmos as well, to help in improving the audio coming out of your device. This can be done by taking the following steps:

v. Go to Settings

vi. Select "Sounds and vibration menu"

vii. Tap on "Advanced sound settings"

viii.Tap "Sound quality and effects"

ix. Tap "Dolby Atmos"

Turn on Face Unlock

The Face Unlock feature is an alternative unlocking method available on the Galaxy A71. Though this is not as secure as the Face ID that is available on the iPhone, but it can be used in getting the lock screen unlocked.

To do this, you are required to take the following steps:

i. Go to Settings

ii. Select "Biometrics and security"

iii. Select "Face recognition"

iv. You will be prompted by on-screen instructions which should be followed to set it up.

v. This feature can also be customized to take you directly to the home screen by getting the "Stay on Lock screen" option toggled off.

vi. It is, however, advisable to turn off faster recognition as well because leaving it on makes the face unlock less secure.

Restart the phone with your voice

• The power button can be set up to get your phone turned on or even to get another app launched. The Bixby can be used to turn off or restart your phone as well.

• This is done by simply, launching Bixby and making use of a variation of command such as "Restart my phone" or "Turn off my phone." Then you will be asked to confirm the command.

Record Your Screen And Yourself

The screen recording feature is available on Galaxy A71 and it can be used to perform various functions such as walking someone through troubleshooting their phone, recording you doing stuff, etc.

If you are making use of this feature for the first time, you are required to take the following steps:

i. Go to Settings

ii. Select "Advanced features"

iii.Select "Screenshots and screen recorder"

iv.Select "Screen recorder settings"

v. You can set if you want the screen recordings to have audio recorded from just your phone, make use of the microphone and the system audio or no audio at all.

vi. You can also set the video quality as well as getting the side of the selfie overlay adjusted when making use of the front-facing camera.

vii. Swipe down from the top of your screen and select a screen recorder from the quick settings list to take a screen recording.

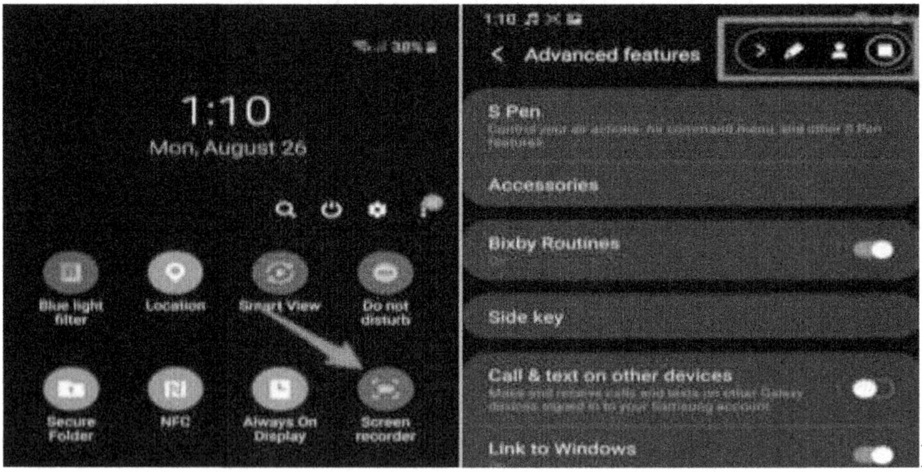

It is to be noted that a toolbar will show up at the top of your screen to help trigger the drawing tool and get the selfie feature turned on or stop recording. Once you are done with your recording, tap on the Stop button.

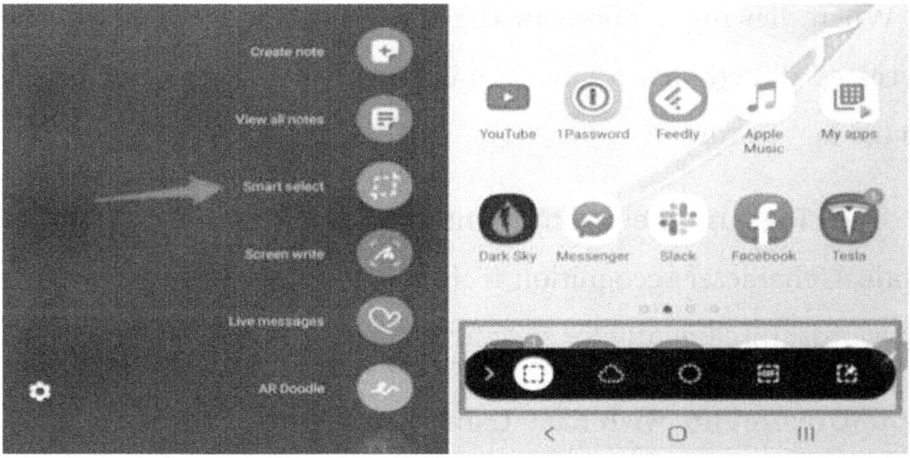

• Go to the toolbar located at the bottom of the screen to pick from a crop tool that makes use of a square or rectangle for your screenshot.

• The cloud is a freehand drawing tool that can be used for an odd-shaped screenshot. The circle makes use of the circle while the GIF records part of your screen to get screen recording created which is then converted into an animated image.

• Lastly, the square icon having the thumbtacks in it helps in taking a screenshot and pins the image to your display.

Effortlessly View Photo Collections with Tag

• The Galaxy A71 upgraded tag experience can be used in viewing collections of similar pictures.

• When viewing a photo in the Gallery, a toolbar that can be accessed directly has a new icon included that takes you directly to the tag feature.

• The Tag also helps in harnessing the Samsung-developed optical character recognition technology which proposes specific keywords intelligently for specific images.

Scan Documents with Ease Using your Phone camera

i. The "Document scan" option can be enabled by taking the following steps:

ii. Go to the Camera settings

iii. Select the Scene optimizer menu as soon as the Camera app is opened which helps in sending documents efficiently with the Galaxy A71.

Make the Edge Panel your own

It is normal to have people that get in touch with you most of the time as well as apps you get to use mostly on your device. Due to this reason, the GalaxyA71 is designed with a customizable edge panel to help you in getting easy and faster access to your favorites.

Drag the tab on the right-hand side of your Galaxy A71 to access the shortcut tray of favorite apps and services. Take the following steps:

i. Go to Settings

ii. Tap "Display"

iii. Tap "Edge screen"

iv. Select "Edge panels"

The contacts and apps that are displayed can be fully customized and also get to set app pairs which include apps that are commonly used along with each other such as the email and calendar, to get launched at the same time.

Get Intelligent To Save Power

- The GalaxyA71 are designed with an intelligent battery that can be used all day long. Irrespective of this feature, there might still be a need to keep your device always powered up especially for the users that make use of the device most of the time.

- To help with this, the GalaxyA71 are designed with Adaptive Power Saving Mode which can be found under the Device care section in the Settings.

- This mode makes use of machine learning in monitoring and learning about you're your device can be used and it also automatically set your device to medium power-saving mode once it detects that you are making use of your phone than usual.

Chapter 34: APPENDIX

About Accessibility

To make the device easy yfor people having medical challenges such as impaired vision reduced dexterity and hearing to use, you can make use of some features in improving the device's accessibility.

i. Go to the Settings app

ii. Tap "Accessibility"

iii. Go to the top of the screen to check the accessibility feature in use.

iv. Tap to see the features that are currently in use or tap "Turn off" to get features disabled quickly.

Voice Assistant (Voice feedback)

Activating or deactivating Voice Assistant

Any time the Voice assistant is activated, voice feedback will be automatically started by the device.

The device will read the text available on the device's screen aloud once you tap the selected feature once while tapping twice will get the feature activated.

i. Open the Settings app

ii. Tap Accessibility

iii. Tap Screen reader

iv. To activate the Voice assistant, tap on its switch.

v. To get the Voice assistant deactivated, tap on the Switch then tap anywhere on the screen twice in quick succession.

vi. Tap Tutorial to view how to make use of the Voice assistant

Note that some features will not be available while making use of the Voice Assistant.

Using the finger gesture to control device screen

In the process of making use of the Voice assistant, different finger gestures can be used in controlling the device's screen.

One Finger Gestures

Tapping: Get your finger placed on the screen and move around to explore on the screen.

With this, you can also get to read items aloud under your finger.

Tapping twice: Get selected item opened. While the selected items are being read aloud, get your finger released from the item once you hear the item you want. Tap anywhere on the screen twice in quick succession afterward.

Double-tapping and holding the screen: This enables you to move an item or gain access to an available option.

Swiping to the left: This option enables you to read previous items

Swiping to the right: This option enables you to read the next item

Swiping upwards or downwards: This enables the most recent contextual menu settings to be changed

Swiping to the left then right in one motion: This helps you in scrolling up the screen

Swiping to the right then left in one motion: This helps you in scrolling down the screen

Swiping upwards then downwards in one motion: With this option, you can get to move to the first item on the screen

Swiping downwards then upwards in one motion: With this option, you will get to move to the last item on the screen

Two-Finger Gestures

Tapping twice: This starts, pause or resume current actions

Tapping three times: The current time, remaining battery power and many more can be checked using this gesture

To get items that are read aloud changed; you are required to take the following settings:

i. Open the Settings app

ii. Tap “Accessibility”

iii. Tap “Screen reader”

iv. Tap “Settings”

v. Select “Status bar information”

Tapping four times: Use this in activating or deactivating voice feedback

Swiping to the left: Use this in moving to the next page

Swiping to the right: Use this in returning to the previous page

Swiping upwards: Use this gesture in scrolling down the list

Swipe downwards: Use this in scrolling up the list.

Swipe upwards or downwards on the Home screen: This gesture helps in opening up the Apps screens

Swipe upwards or downwards on the Apps screen: This helps in returning to the Home screen

Swiping in any direction on the locked screen: This gesture helps in getting your screen unlocked.

Swiping downwards from the top of the screen: This gesture helps in opening the notification panel

Three-Finger Gestures

Tapping: You use this to start reading items aloud from the top.

Tapping twice: With this gesture, reading can start from the next item.

Tapping three times: Use this in reading the last selected text and get it copied to the clipboard.

Swiping to the left or right: Use this gesture in opening the contextual menu and scrolling through the options

Swiping upwards or downwards: Use this in changing the text reading and selection of granularity

Swiping upwards then downwards in one motion: This can be used in returning to the previous screen

Swiping downwards then upwards in one motion: Use this gesture in returning to the Home screen

Four Finger Gestures

Tapping: Use this gesture in returning to the previous screen

Tapping twice: Use this gesture in returning to the Home screen

Tapping three times: Use this gesture in opening recent apps' list.

Using the Contextual Menu

- Swipe to the left or right using three fingers when making use of the Voice assistant to open up the contextual menu which causes the device to read options aloud.

- Swipe to the left or right making use of three fingers for scrolling through the options.

- Swipe upwards or downwards to make use of the option you want once you hear it or get the settings of the option-adjusted. An example of this is when you hear the speech rate; you can swipe upward or downward to get it adjusted.

Adding and Managing Image Labels

Labels can be assigned to images on the screen. Once the labels are assigned, they are read aloud by the device once the images are selected.

Tap on the screen twice making use of three fingers and hold to get labels added to images.

Take the following steps to manage the labels:

i. Open the Settings app

ii. Tap "Accessibility"

iii. Tap "Screen reader"

iv. Select "Settings"

v. Select "Manage custom labels"

Configuring Settings for Voice Assistant

It is advisable to get your Voice assistant configured for convenience. You can do this by taking the following steps:

i. Open the Settings app

ii. Tap "Accessibility"

iii. Tap "Screen reader"

iv. Select "Settings"

Text-to-speech: When the voice assistant is activated, text-to-speech features' settings are altered. These features include languages, speeds, etc.

Verbosity: Gets the detailed settings customized for voice feedback.

Shake for continuous reading: The device can be set to read the text displayed on the screen aloud once the device is shaken. You will also have an option to choose from different shaking intensity.

Speak passwords: The device can be set to read characters aloud as well as when a password is entered.

Reader caller ID aloud: Your device can be set to read the caller's name aloud when there is an incoming call.

Vibration feedback: The device can be set to vibrate when you get the screen controlled such as the selected of an item.

Sound feedback: The device can be set to emit a sound when controlling the screen such as in the process of selecting an item. The volume of the sound played can also be adjusted when you touch the screen to control it.

Focus on speech audio: The device can be set to get the media volume turned down when the device reads an item aloud.

Focus indicator: The device can be set to thicken the border of the focus indicator which appears when items are tapped. The color of the indicator can also be changed.

Single-tap activation: You can set the device such that a single tap will activate the selected item.

Single-tap navigation bar: The device can be set such that you will get to make use of buttons available on the navigation bar with a single tap.

Quick menu: The contextual menu options can be selected to display whenever you swipe to the right or left with the use of your three fingers.

Manage custom labels: Added labels can be managed using this option.

Keyboard shortcuts: When making use of an external keyboard, you can set the key shortcuts to use.

Dark screen with Side key: While the screen is off, you can set the device to function normally.

Status bar information: Here you can select the items that will be read aloud when you tap the screen three times with the use of two fingers.

Developer options: Here, you will get to set options for app development.

Making Use of the Device with the Screen Turned Off

You can configure your device to function normally even with the screen off for privacy reasons. Tapping on the screen or pressing keys will not turn on the screen.

Note that when the screen is off, the currently activated features will not be turned off.

i. Open the Settings app

ii. Select "Accessibility"

iii. Tap "Screen reader"

iv. Tap "Settings"

v. Tap the Dark screen with Side key switch to get it activated. Note that you can get this feature activated or deactivated by pressing the side key twice quickly.

You should also note that the feature will only appear when the Voice assistant has been activated.

Entering Text Using the Keyboard

To get the keyboard displayed:

i. Tap the text input field

ii.Tap anywhere on the screen twice in quick succession

To activate the rapid key input feature:

i. Open the Settings app

ii. Tap "Accessibility"

iii. Select "Screen reader"

iv. Tap on the Quick typing switch to get it activated

Whenever you make use of your finger in touching the keyboard, the device will read the characters under your finger aloud. Once you hear the character you want, get your finger released from the screen to get it selected. The character will be entered and the text will be read aloud by your device.

If the Quick typing is not activated, get your fingers released from the character that you want and tap anywhere on the screen twice in quick succession. But ensure to activate the Voice assistant for this feature to be activated.

Entering Additional Characters

- To input additional characters that are available for the key, tap and hold a key on the keyboard.
- You will then be prompted with a window above the key which shows the available characters.
- For the selection of a character, drag your finger on the pop-up window till you hear the character you want and release afterward.

Visibility Enhancements Using the High Contrast Theme

i. Open the Settings app

ii. Tap "Accessibility"

iii. Tap "Visibility enhancements"

iv. Tap "High contrast theme"

v. Select your preferred theme

Making Use of the High Contrast Font

The color and outline of fonts can be adjusted to help in increasing the contrast between the text and background by taking the following steps:

i. Open the Settings app

ii. Tap “Accessibility”

iii.Tap “Visibility enhancements”

iv. Tap the High contrast fonts switch to get it activated.

Making Use of the High Contrast Keyboard

The color of the Samsung keyboard can be changed to help in increasing the contrast between the text and the background.

i. Open the Settings app

ii. Tap “Accessibility”

iii. Tap “Visibility enhancements”

iv. Tap the High contrast keyboard switch to get it activated.

Displaying Button Shapes

By getting button shapes displayed with frames helps them in standing out and you can do this by taking the following steps:

i. Open the Settings app

ii. Tap “Accessibility”

iii. Tap “Visibility enhancements”

iv. Tap the Show button shapes switch to get it activated.

Reversing the Display Colors

Doing this will help in improving screen visibility which helps users in recognizing the text on the screen easily. To do this, take the following steps:

i. Open the Settings app

ii. Tap “Accessibility”

iii. Tap “Visibility enhancements”

iv. Tap the Negative colors switch to get it activated.

Color adjustment

This feature helps to adjust how colors are displayed on your device screen if there are issues in identifying colors. The device does this by changing the colors into more recognizable colors.

i. Open the Settings app

ii. Tap “Accessibility”

iii. Tap “Visibility enhancements”

iv. Tap "Color adjustment"

v. Tap on the switch to get it activated.

vi. Select an option from the available ones.

vii. Drag the adjustment bar under intensity to get color intensity adjusted.

viii. Selecting a personalized color helps you to get the color of the screen personalized. You will be prompted by on-screen instructions to help complete the color adjustment process.

Color Lens

If you have difficulties reading the text, the screen colors can be adjusted to solve this.

i. Open the Settings app

ii. Tap "Accessibility"

iii. Tap "Visibility enhancements"

iv. Tap "Color lens"

v. Tap on the switch to get it activated.

vi. Select your preferred color

vii. Adjust the transparency level by dragging the adjustment bar under Opacity.

Reducing Screen Motion Effects

In a situation where you are sensitive to animations or screen movement, you can reduce the animation effects by taking the following steps:

i. Open the Settings app

ii. Tap "Accessibility"

iii. Tap "Visibility enhancements"

iii. Tap the remove animations switch to get it activated.

Making use of the magnifier

Your device screen can get magnified and zoomed in on a specific area by taking the following steps:

i. Open the Settings app

ii. Tap "Accessibility"

iii. Tap "Visibility enhancements"

iv. Select "Magnification"

• Triple tap screen to magnify: This gesture allows you to zoom in by tapping the screen thrice. Tapping the screen three times again will return you to your previous or normal view.

• Tap button to magnify: Using this enables you to zoom in by tapping on the navigation bar as well as tapping wherever you like to magnify on the screen. You can tap to return to the normal view.

• By dragging two or more fingers across the magnified screen, you will get to explore the screen.

• Pinch two or more fingers on the magnified screen or spread them apart to adjust the zoom ratio.

• Tapping the screen thrice and holding it helps you to magnify your screen temporarily. While you are holding the screen, you can drag your finger to explore the screen and by releasing your finger, you will be returned to the normal view.

NOTE

• It is important to note that you can't magnify the keyboards as well as the navigation bar on the screen.

• The performance of some apps will also be affected when the feature is activated.

Enlarging Mouse or Touchpad Pointers

This can be done by taking the following steps:

i. Open the Settings app

ii. Tap "Accessibility"

iii. Tap "Visibility enhancements"

iv. Tap the Large mouse/touchpad pointer switch to get it activated

How to Change the Font

You can change your font size and style by taking the following steps:

i. Open the Settings app

ii. Tap "Accessibility"

iii. Tap "Visibility enhancements"

iv. Tap "Font size and style"

How to Change The Screen Zoom Setting

To do this, you are required to take the following steps:

i. Open the Settings app

ii. Tap “Accessibility”

iii. Tap “Visibility enhancements”

iv. Tap “Screen zoom”

Hearing Enhancements

Sound detectors

- Your device can be configured to detect if a baby is crying or there is a sound of your doorbell.

Once any of these sounds are detected, your device will vibrate while also saving the alert as a log.

- It is importan to note that voice recognition will not work while you have this feature enabled.

Baby crying detector

i. Open the Settings app

ii. Tap “Accessibility”

iii. Tap “Hearing enhancements”

iv. Select “Sound detectors”

v. Tap “Baby crying detector”

vi. Tap "Start"

Doorbell detector

i. Open the Settings app

ii. Tap "Accessibility"

iii. Tap "Hearing enhancements"

iv. Select "Sound detectors"

v. Tap "Doorbell detector"

vi. To make a recording of your doorbell, select the "Record option." This will enable the device to record your doorbell and detect it easily.

vii. To ensure your doorbell has been correctly recorded, tap the "Test and ring your doorbell" option.

viii. Once the device can detect the sound of your doorbell, select "Start."

ix. Tap on "Change the doorbell sound" to change the sound of your doorbell

Turning Off All Sounds

Your device can be configured such that all device sounds such as your media sounds including the voice of the caller during a call, will be muted. To do this:

i. Open the Settings app

ii. Tap "Accessibility"

iii. Tap "Hearing enhancements"

iv. Tap the Mute all sounds switch to get it activated.

Hearing Aid Support

Your device can be configured to improve your sound quality such that it works much better when using your hearing aids by taking the following steps:

i. Open the Settings app

ii. Tap "Accessibility"

iii. Tap "Hearing enhancements"

iv. Tap the Hearing aid support switch to get it activated.

How to Adjust The Sound Balance

You can set your device to get the sound balance adjusted when making use of an earphone by following the instructions below.

i. Get the earphones connected to the device and wear it.

ii. Open the Settings app

iii. Tap "Accessibility"

iv. Tap "Hearing enhancements"

v. Then get the adjustment bar dragged under the Left/right sound balance to the left and right and get the sound balance adjusted.

Mono Audio

The stereo sound that is played through all earphone speakers are combined into one signal by the mono output.

It is recommended for you to make use of this if you have a hearing impairment or if you are convenient making use of a single earbud.

i. Open the Settings app

ii. Tap "Accessibility"

iii. Tap "Hearing enhancements"

iv. Tao the Mono audio switch to get it activated

Caption settings

i. Open the Settings app

ii. Tap "Accessibility"

iii. Tap "Hearing enhancements"

iv. Select Samsung subtitles (CC) or Google subtitles (CC)

v. Tap the switch to get it activated.

vi. Choose an option for getting your caption settings configured.

Speech-To-Text

The voice can be simultaneously recorded and converted to on-screen text by your device when you make use of this feature.

i. Open the Settings app

ii. Tap "Accessibility"

iii. Tap "Hearing enhancements"

iv. Tap "Speech-to-text"

v. Tap and make use of your microphone to make recordings

vi. Tap DONE, when you have finished recording

It is to be noted that you can open the Voice recorder app to access recorded files or alter the voice memo system language.

Interaction and Dexterity Universal Switch

The touch screen of your device can be controlled by getting an external switch connected to it, by tapping the screen or by making use of head movements as well as face gestures. Take the following steps to activate the feature:

i. Open the Settings app

ii. Tap "Accessibility"

iii. Tap "Interaction and dexterity"

iv. Select "Universal switch"

v. You will be prompted with on-screen instructions to read.

vi. Tap DONE

vii. Tap the switch to get it activated

viii. In a situation where the universal switches are not registered on the device, tap "Add switch," then set the switches to control the device afterward.

NOTE:

- At least one switch must be registered on the device to get the universal switch feature activated.

- Tap on the switch under “Switches” to set the switches to control the device.

- Press the Side key as well as the Volume Up key simultaneously to get this feature deactivated.

Assistant Menu

- To access the assistive menus more easily from any screen, they have the assistive shortcut icon displayed as a floating icon.

- As soon as the assistive shortcut icon is tapped, the icon will expand slightly thereby displaying the assistive menus in the icon.

- To move the other panels, tap the left or right arrow. You can also select other menus by swiping to the left or right.

Using the Cursor

- Tap the cursor on the assistive menu.

- The screen can be controlled by making use of small finger movements on the touch area.

- To move the cursor, all you need to do is dragging your finger on the touch area.

- To get items under the cursor selected, tap the screen.

You can make use of the following options as well:

- Scroll left or right on the screen.
- Scroll up or down the screen.
- Magnify the area where the cursor is located.
- Close the touch area.

Using the Enhance Assistive Menus

The device can be configured to display enhanced assistive menus for selected apps by taking the following steps:

i. Open the Settings app

ii. Tap "Accessibility"

iii. Tap "Interaction and dexterity"

iv. Tap "Assistant menu"

v. Tap the "Assistant plus"

vi. Tap the switches that are next to the apps you want to get activated.

Click After Pointer Stops

The device can be configured to automatically get items selected when the mouse pointer is placed over the item

i. Open the Settings app

ii. Tap "Accessibility"

iii. Tap "Interaction and dexterity"

iv. Tap the Click after pointer stops switch to get it activated.

Answering or ending calls

i. Open the Settings app

ii. Tap "Accessibility"

iii. Tap "Interaction and dexterity"

iv. Tap "Answering and ending calls"

v. Tap the switch of your preferred method.

Interaction Control

To help in restricting your device's reaction to inputs while making use of apps, you can get the interaction control mode activated.

i. Open the Settings app

ii. Tap "Accessibility"

iii. Tap "Interaction and dexterity"

iv. Tap "Interaction control"

v. Tap the switch to get it activated.

vi. Press and hold the Side key as well as the Volume Up key simultaneously while the app is in use.

vii. Get the size of the frame adjusted or get a line drawn around an area you like to restrict.

viii. Select "Done"

NOTE:

- The restricted area will be displayed by the device.
- There will be no reaction when the restricted area is touched and the device's keys will get disabled.

• The functions of the Side key, Volume key as well as the keyboard can be enabled in the interaction control mode.

• Press and hold the Side key with the Volume Up key simultaneously to help in deactivating the interaction control mode.

• The device can be configured to automatically lock the screen after the interaction control mode has been deactivated by taking the following steps:

i. Open the Settings app

ii. Tap "Accessibility"

iii. Tap "Interaction and dexterity"

iv. Tap "Interaction control"

v. Tap the Lock screen when deactivated switch to get it activated.

Touch and Hold Delay

You can configure a recognition time for tapping and holding the screen by taking the following steps:

i. Open the Settings app

ii. Tap "Accessibility"

iii. Tap “Interaction and dexterity”

iv. Select “Touch and hold delay”

v. You will be prompted with options from which you can select from.

vi. Tap duration

The duration of time you must touch the screen before the device can recognize it as a tap can also be configured by taking the following steps:

i. Open the Settings app

ii. Tap “Accessibility”

iii. Tap “Interaction and dexterity”

iv. Tap duration

v. Tap on the switch to get it activated

vi. Set the time

Ignore Repeated Touches

The device can be configured to recognize the only first tap within a set of times when the screen is tapped repeatedly.

i. Open the Settings app

ii. Tap “Accessibility”

iii. Tap “Interaction and dexterity”

iv. Tap “Ignore repeated touches” and tap the switch to get it activated

v. Set the time

Advanced Settings

Launching the accessibility features quickly

It is recommended to set your device to launch accessibility features quickly by taking the following steps:

i. Open the Settings app

ii. Tap “Accessibility”

iii. Tap “Advanced settings”

iv. Select “Side and Volume up keys”

v. Tap the switch to get it activated.

vi. Choose an accessibility feature that helps in the opening when you simultaneously hit the Side key and Volume Up key.

The following accessibility features can be accessed:

- Accessibility
- Negative colors
- Universal switch
- Color adjustment
- Voice Assistant
- Color lens
- Magnifier window
- Interaction control

Launching the Shortcut Services Quickly

You can configure your device to start a shortcut service quickly by taking the following steps:

i. Open the Settings app

ii. Tap "Accessibility"

iii. Tap "Advanced settings"

iv. Tap "Volume up and down keys"

v.Tap the switch to get it activated.

vi. Tap on "Selected service" and choose a shortcut service to start when you press and hold the Volume Up key and Volume down key simultaneously for three seconds.

It is to be noted that the following shortcut services are accessible:

- Your phone companion
- Universal switch
- Voice assistant

Setting light notification

The device can be configured to show notifications such as incoming calls or new messages by blinking the flash. To do this, you are required to take the following steps:

i. Open the Settings app

ii. Tap "Accessibility"

iii.Tap "Advanced settings"

iv. Select "Flash notification"

v. Tap the switches next to the options you like to activate

Notification Reminders

Your device can be configured to alert you of notifications that you have forgotten to check by taking the following steps:

i. Open the Settings app

ii. Tap "Accessibility"

iii. Tap "Advanced settings"

iv. Tap Notification reminders

v. Tap the switch to get it activated

The following options can be selected:

Vibrate when sound plays: The device is configured to vibrate and play a notification sound when there is the presence of unchecked notifications.

Remind every: Configures an interval between alerts.

Selected reminders: Configures the apps such that it alerts you of notifications.

Adding Voice Recordings to Voice Labels

- Voice labels can be used in distinguishing objects having similar shapes by getting labels attached to them. A voice recording can be recorded and assigned to an NFC-enabled voice label.

• When your device is placed near to the label, the voice recording gets played back.

To make use of this feature, you will need to have the NFC feature turned on by taking the following steps:

i. Open the Settings app

ii. Tap “Accessibility”

iii. Tap “Advanced settings”

iv.Tap “Voice label” and the voice recorder will be launched.

v. Select “Start recording” and speak into the microphone.

vi. Tap DONE to stop when you are done recording

vii. Hold the device back over the voice label and the voice information will be written to the voice label.

Installed services

This feature enables you to access the accessibility services that are installed on the device.

i. Open the Settings app

ii. Tap “Accessibility”

iv. Tap “Installed services”

Conclusion

With these tips, we believe you can now start using your Samsung Galaxy A71 like a Pro. If you do find this guide useful, please do drop a review on Amazon so we get to know your thought.

Best Regards,

Sam J. Wealth

Made in the USA
Middletown, DE
22 August 2021

46683939R00146